WORLD BANK WORKING PAPER NO. 150

# Globalization and Technology Absorption in Europe and Central Asia

*The Role of Trade, FDI, and Cross-border Knowledge Flows*

*Itzhak Goldberg*
*Lee Branstetter*
*John Gabriel Goddard*
*Smita Kuriakose*

THE WORLD BANK
Washington, D.C.

LIBRARY
GRANT MacEWAN
COLLEGE

Copyright © 2008
The International Bank for Reconstruction and Development/The World Bank
1818 H Street, N.W.
Washington, D.C. 20433, U.S.A.
All rights reserved
Manufactured in the United States of America
First Printing: June 2008

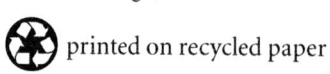

♻ printed on recycled paper

1 2 3 4 5 11 10 09 08

World Bank Working Papers are published to communicate the results of the Bank's work to the development community with the least possible delay. The manuscript of this paper therefore has not been prepared in accordance with the procedures appropriate to formally-edited texts. Some sources cited in this paper may be informal documents that are not readily available.

The findings, interpretations, and conclusions expressed herein are those of the author(s) and do not necessarily reflect the views of the International Bank for Reconstruction and Development/The World Bank and its affiliated organizations, or those of the Executive Directors of The World Bank or the governments they represent.

The World Bank does not guarantee the accuracy of the data included in this work. The boundaries, colors, denominations, and other information shown on any map in this work do not imply any judgment on the part of The World Bank of the legal status of any territory or the endorsement or acceptance of such boundaries.

The material in this publication is copyrighted. Copying and/or transmitting portions or all of this work without permission may be a violation of applicable law. The International Bank for Reconstruction and Development/The World Bank encourages dissemination of its work and will normally grant permission promptly to reproduce portions of the work.

For permission to photocopy or reprint any part of this work, please send a request with complete information to the Copyright Clearance Center, Inc., 222 Rosewood Drive, Danvers, MA 01923, USA, Tel: 978-750-8400, Fax: 978-750-4470, www.copyright.com.

All other queries on rights and licenses, including subsidiary rights, should be addressed to the Office of the Publisher, The World Bank, 1818 H Street NW, Washington, DC 20433, USA, Fax: 202-522-2422, email: pubrights@worldbank.org.

ISBN-13: 978-0-8213-7583-9
eISBN: 978-0-8213-7584-6
ISSN: 1726-5878        DOI: 10.1596/978-0-8213-7583-9

Cover image by Esteban Montes.

**Library of Congress Cataloging-in-Publication Data.**

Globalization and technology absorption in Europe and Central Asia : the role of trade, FDI, and cross-border knowledge flows / Itzhak Goldberg ... [et al.].
  p. cm. -- (World Bank working paper ; no. 150)
 Includes bibliographical references and index.
 ISBN 978-0-8213-7583-9 -- ISBN 978-0-8213-7584-6 (electronic)
 1. Technology transfer--Asia, Central. 2. Technology transfer--Europe. 3. Investments, Foreign--Asia, Central. 4. Investments, Foreign--Europe. 5. Globalization--Asia, Central. 6. Globalization--Europe. I. Goldberg, Itzhak.
 HC420.3.Z9T44 2008
  303.48'33094--dc22

2008018986

# Contents

Foreword .......................................................... vii

Acknowledgments ................................................... ix

Executive Summary ................................................. xi

1. **Introduction** .................................................. 1
   Definition of Innovation and Knowledge Absorption ................ 1
   Diffusion of Knowledge in Support of Productivity Growth: Literature Review ... 2
   Conceptual Framework: Economic Conditions and Capacities
   for Knowledge Absorption ......................................... 4

2. **Patents as Indicators of Technological Activity in the ECA Region** ... 7
   Patent Data Provide a View of the Knowledge Absorption Process ... 7
   Implications for Policy ......................................... 18

3. **The Links among Knowledge Absorption, Trade, and FDI** ........ 23
   Trade and its Benefits .......................................... 24
   Foreign Direct Investment ....................................... 31
   Discussion of Results with Implications for Policy .............. 49

4. **How Does FDI via Company Acquisition Impact Technology Absorption?
   A Case Study of Serbian Enterprises** ........................... 55
   Investment Climate and Sequence of Mergers and Acqusitions,
   and Greenfield FDI .............................................. 57
   The Background of the Serbian Privatization Program ............. 61
   Policy Implications from Case Studies ........................... 76

Appendixes ........................................................ 81
   A. Statistical Tables for Chapter 2 ............................. 83
   B. Regression Variables Used in Chapter 3 ...................... 85
   C. Questionnaire for Company Interviews ........................ 87
   D. Correlates of ICT and Quality Certification ................. 93

Bibliography ..................................................... 115

### LIST OF TABLES

1. Correlations of "Composite" Measure of Absorption with Firm
   Characteristics [+] positive correlation, [−] negative correlation ... xix

2. Comparing the Effects of Privatization on FDI versus
Domestically-owned Firms................................................. xxi
2.1. Top 10 Russian Generators of U.S. Patents ................................ 15
3.1. Trade Restrictiveness Indices............................................. 27
3.2. Logistics Performance Indices, 2007..................................... 30
3.3. Regression Results for New Product Introductions as Dependent Variable ...... 39
3.4. Regression Results for Product Upgrades as Dependent Variable .............. 42
3.5. Regression Results for Introduction of New Technology
as Dependent Variable .................................................. 44
3.6. Linear Regression Results Based on a "Composite" Measure of Absorption...... 46
3.7. Regression Results Based on Composite Measure of Absorption, Panel Data .... 48
4.1. Results of the Serbian Privatization Program............................... 62
4.2. Revenue and Employment Trends Pre- and Post-acquisition .................. 65
4.3. Productivity Trends Pre- and Post-acquisition ............................. 67
4.4. Financial Ratios Pre- and Post-acquisition................................. 69
A.1. Patent Citations in ECA and Comparator Regions .......................... 83
A.2. Hypothesis Tests for Equality of Sample Means............................ 84
B.1. Definition of Variables Used in Regressions............................... 85
D.1. ISO Certification and Web Use across Sectors............................. 97
D.2. ISO Certification and Web Use, Firm Size, and Age ........................ 98
D.3. ISO Certification and Web Use, Firm Ownership, and Trade Integration........ 98
D.4. Variable Definitions .................................................... 102
D.5. Determinants of ISO Certification and Web Use—Cross-Sectional
Regressions .......................................................... 104
D.6. Determinants of Technology Adoption across Country Groups .............. 106
D.7. Determinants of ISO Certification and Web-Use—Panel Regressions ......... 108

## List of Figures

1. U.S. Patents Granted per Million Population ............................... xiii
2. U.S. Patent Grants for the ECA 7 vs. India and China ....................... xiv
3. Indigenous Patents and Coinventions in ECA: 1993–2007.................... xv
4. Trade Restrictiveness Index (TRI), 2005–06................................ xvii
5. FDI Inflows as a Percentage of GDP, 2005................................. xviii
1.1. Innovation and Absorption as Inputs into Growth and Productivity............ 5
2.1. ECA Region Patenting in the EPO ....................................... 9
2.2. ECA Patenting in Europe, by First Inventor Country of Residence............. 10
2.3. ECA Coinvention by Partner Country .................................... 11

2.4. U.S. Patent Grants for the ECA 7, India, and China . . . . . . . . . . . . . . . . . . . . . . . . 12
2.5. The Expanding Role of International Coinvention in the ECA 7. . . . . . . . . . . . . . 13
2.6. International Coinvention in Russia. . . . . . . . . . . . . . . . . . . . . . . . . . . . . . . . . . . . 14
2.7. National Innovation Systems. . . . . . . . . . . . . . . . . . . . . . . . . . . . . . . . . . . . . . . . . 21
3.1. Overall Trade Restrictiveness Index-tariff (all goods), 2006. . . . . . . . . . . . . . . . . 28
3.2. Overall Trade Restrictiveness Index-tariff + non-tariff (all goods), 2006 . . . . . . . 28
3.3. . . . . . . . . . . . . . . . . . . . . . . . . . . . . . . . . . . . . . . . . . . . . . . . . . . . . . . . . . . . . . . . . . 35
3.4. . . . . . . . . . . . . . . . . . . . . . . . . . . . . . . . . . . . . . . . . . . . . . . . . . . . . . . . . . . . . . . . . . 35
D.1. ISO Certification and Web Use across ECA Countries . . . . . . . . . . . . . . . . . . . . 96

## LIST OF BOXES

3.1. Measuring the Impact of Trade and FDI on Technology Absorption . . . . . . . . . . 34
3.2. New Foreign Competition: Lowering Profits and Raising Efficiency. . . . . . . . . . . 35
4.1. Methodology: Company Selection, Data Sources, Fieldwork. . . . . . . . . . . . . . . . 63
4.2. A Closer Look at Productivity Trends. . . . . . . . . . . . . . . . . . . . . . . . . . . . . . . . . . . 68
D.1. Identification Strategy . . . . . . . . . . . . . . . . . . . . . . . . . . . . . . . . . . . . . . . . . . . . . . 101

# Foreword

This report on globalization and technology absorption in Europe and Central Asia is part of the Europe and Central Asia Knowledge Economy Flagship Studies produced by the Finance and Private Sector Development Department. Innovation and cross-border absorption of knowledge are central forces behind economic convergence and a more sustained knowledge-intensive growth. Absorption of technology is considered a necessary step to promote the development of human capital and the productive base, paving the way for innovations at the global knowledge frontier. Research and development, patents, trade, and foreign direct investment are major channels of technological absorption, allowing diffusion of new ideas and manufacturing best practices among countries and firms. These channels constitute the central focus of this study, which is the second publication in this Knowledge Economy series. The first title in the series examined the public financial support of commercial innovation. The third title of the series aims to address the importance of the restructuring and/or exit of existing public R&D institutions, and to provide policy lessons on Research and Development Institutes restructuring.

This study uses patent databases, surveys of enterprises, and case studies to investigate how the presence of specific channels of absorption molds decision making about technology at the firm level. Trade and FDI flows show considerable promise as catalysts for the region to upgrade its technology and move near the global technology frontier. Econometric analysis using enterprise surveys from all ECA countries helps us understand the conditions and policies that induce firms to incorporate external knowledge and technology into their overall growth strategies.

We find evidence of learning by exporting, underscoring the importance of trade as a driver of technology absorption. A case study of several manufacturing firms in Serbia complements this perspective, providing a detailed picture of the positive dynamics that are produced by FDI, particularly in terms of investment and risk-taking incentives that are critical for technology absorption. Examination of patent citations shows that cross-border knowledge flows remain weak. Consequently, science and innovation policies should foster greater integration of the region's substantial science and engineering resources with those of the rest of the world. These policies should encourage international collaboration and closer connection between the region's public sector R&D and global private sector efforts.

**Fernando Montes-Negret**
*Director, Private and Financial Sector Development*
*The World Bank*

# Acknowledgments

This Knowledge Economy Study is part of an ongoing Regional Working Paper Series sponsored by the Chief Economist's Office in Europe and Central Asia Region of the World Bank. It was financed by the Europe and Central Asia Regional Studies Program.

This study was prepared by a team consisting of Itzhak Goldberg (World Bank, Team Leader), Lee Branstetter (Professor, Heinz School, Carnegie Mellon University), John Gabriel Goddard, and Smita Kuriakose (both World Bank). Significant contributions were received from Martina Kobilicova, Andrej Popovic, Lazar Sestovic, and Jasna Vukoje (all World Bank) for the Serbian case studies. Contributions were also received from Ana Margarida Fernandes, Paulo Guilherme Correa, Mallika Shakya, and Chris Uregian (all World Bank).

We would like to acknowledge valuable comments and suggestions from David Tarr (World Bank), Holger Görg (Professor, Department of Economics, University of Nottingham), Brett Coleman, Jean Louis Racine, Donato De Rosa, and Joanna Tobiason (all World Bank). Peer reviewers included Professor Howard Pack (Wharton Business School, University of Pennsylvania), Milan Brahmbhatt, Paloma Anos Casero, and Mark Dutz (all World Bank).

We would like to especially thank Fernando Montes-Negret, Director of the Private and Financial Sector Development Department, Europe and Central Asia Region, the World Bank, for all his support, comments, and guidance for this study. We would also like to thank Marianne Fay, Lead Economist, in the Europe and Central Asia Chief Economist's Office and Pradeep Mitra, Chief Economist, Europe and Central Asia for their intellectual guidance.

# Executive Summary

Improving the absorptive technology capability of countries—their ability to tap into the global technology pool—is an important mechanism for accelerating industrial development, raising productivity of workers, and increasing economic growth. Trade flows, foreign direct investment (FDI), research and development (R&D), and labor mobility and training are widely accepted as key mechanisms for knowledge absorption. Furthermore, the wealth of detailed information that patents and patent citations contain offer a useful window into the technological absorption process in Europe and Central Asia (ECA). While patents are indications of new-to-the-world innovation, much of this innovation is incremental, building closely on technical foundations developed in foreign countries. Patent *citations* connect ECA inventions to the prior foreign inventions upon which ECA inventions build, tracing pathways of international knowledge diffusion.

The process of knowledge absorption is neither automatic nor costless. It requires a favorable investment climate, as well good national education and research and development systems. This study analyzes the extent of knowledge and technology absorption for firms in ECA, as well as the factors that influence absorption, using statistical analyses of various data sources, including the World Bank Enterprise Surveys, patent databases maintained by the U.S. and European patent offices, and case studies.

The study addresses the following issues:

- What can we learn from patents and patent citations about international knowledge flows and cross-national technological cooperation in ECA? *(Chapter 2)*
- How does openness to trade, participation in global supply networks, and investment in human capital, via on-the-job training, enhance knowledge and technology absorption in ECA-region manufacturing firms? *(Chapter 3)*
- How does FDI stimulate acquisition of managerial and technical skills, new machinery and equipment, and market development? *(Chapter 4)*

## Definitions and Framework

Absorption is a costly learning activity that a firm can employ to integrate and commercialize knowledge and technology that is new to the firm, but not new to the world. For simplicity, development of new-to-the-world knowledge can be considered innovation. In other words, innovation shifts a notional technological frontier outward, while absorption moves the firm closer to the frontier. Examples of absorption include: adopting new products and manufacturing processes developed elsewhere, upgrading old products and processes, licensing technology, improving organizational efficiency, and achieving quality certification. The text box below defines some key terms that refer to absorption and innovation.

This Europe and Central Asia Knowledge Economy Study Part II (ECAKE II) is a followup to the study on *Public Financial Support of Commercial Innovation* (ECAKE I),[1] which

---

1. Available at: http://imagebank.worldbank.org/servlet/WDS_IBank_Servlet?pcont=details&menuPK=64154159&searchMenuPK=64154240&theSitePK=501889&eid=000012009_20060504143652&siteName=IMAGEBANK.

> **Definitions**
>
> Absorption versus Innovation: *New to the Firm versus New to the World*
>
> *Absorptive capacity:* A firm's capacity to assess the value of external knowledge and technology, and make necessary investments and organizational changes to absorb and apply this in its productive activities.
>
> *Examples of absorption:* Adoption of a new product or process; upgrading of an old product or process; utilization of a technology license.
>
> *Product innovation:* Development of new products representing discrete improvements over existing ones.
>
> *Process innovation:* Redesign of products or services; "soft innovation," e.g. reorganization of layouts, transport modes, management, and human resources.
>
> *Incremental innovation:* Innovation that builds very closely on technological antecedents and does not involve much technological improvement upon them.

focused on innovation. ECAKE II provides a more detailed analysis of the complementarities between innovation and absorption, with a focus on absorption. As will be shown in our empirical findings, R&D, a key input for innovation, is an input into absorption as well. Indeed, there are important complementarities between innovation and absorptive capacity. Innovation promotes absorptive capacity because the generation of human capital and new ideas, and the associated knowledge spillover effects, help build absorptive capacity. Conversely, the absorption of cutting-edge technology inspires new ideas and innovations.

The conceptual framework that shapes our analysis follows the endogenous production-function approach. Its premise is that innovation and absorption of knowledge, which are central forces behind economic growth, are in turn determined by economic conditions and policies. We consider trade, FDI, R&D, and patents as "channels of absorption," providing conduits for diffusion of knowledge between countries and absorption within firms. These channels constitute the central focus of the study. Our analysis investigates how differing degrees of exposure to international best practices through these channels affects absorption outcomes. A second line of inquiry asks about the conditions and policies that induce firms to make use of these channels as part of their overall growth strategy.

Properly designed economic policies can significantly influence the degree to which a country absorbs new technology, as well as the decisions by firms to undertake investments that do so. The channels of technology absorption—trade, FDI, R&D—need a stable and conducive policy framework and a business-friendly investment climate. At the same time, a firm's ability to absorb this technology and knowledge depends on its organization and the skills of its workforce.

## Patent Citations, International Coinvention, and Multinational Sponsorship of Local Invention

We analyze international flows of disembodied knowledge—know-how other than that represented by physical capital—by analyzing two of their many channels, international coinvention and multinational sponsorship of local inventions. We also utilize patent citations as an indicator of disembodied knowledge flows, and their usefulness for this purpose as demonstrated in the academic literature. While patents are viewed as a form of intellectual property right over some of the economy's innovative outputs, patent citations provide us with a convenient metric of how existing "pieces" of patented knowledge have contributed to the creation and appropriation of new patented knowledge. Innovators aim to invent new products or processes that lead to direct private and social benefits, but in doing so, they may create knowledge that cannot be retained as a trade secret. This knowledge becomes available to other innovators and reduces future R&D costs to all. A company that obtains a patent on an invention is *legally required* to disclose important information about the new technology, and this information becomes public. There is an extensive literature that shows that patents serve as an important channel of technological diffusion, and that patent citations are a good proxy for the actual flows of technological knowledge.

The data on patent citations from the European Patent Office (EPO) and the U.S. Patent and Trademark Office (USPTO) are used as proxies for the knowledge absorption process in order to study flows of knowledge within ECA, and between ECA and the rest of the world. The analysis explores the knowledge flows and "spillovers" present by tracing the citation pathways that link ECA inventions to prior inventions created in ECA and elsewhere in the world. The extent to which ECA inventors cite new technologies may indicate the extent to which inventors are grounded in the recent state of the art. This study also explores the patterns among cross-national teams of inventors. Patents created by teams that include both ECA-based inventors and those in other countries can reflect the extent to which ECA inventors are connected to the global technological mainstream. While openness to trade and foreign investment allows firms to tap into and benefit from the global pool of knowledge, patent coinventions can also help build these global linkages, and in turn, create knowledge spillovers.

Within ECA, there are differences across countries, with four clear leaders: Hungary, the Czech Republic, the Russian Federation, and Poland. Among these four, Hungary and the Czech Republic fare significantly better than Russia and Poland (Figure 1).

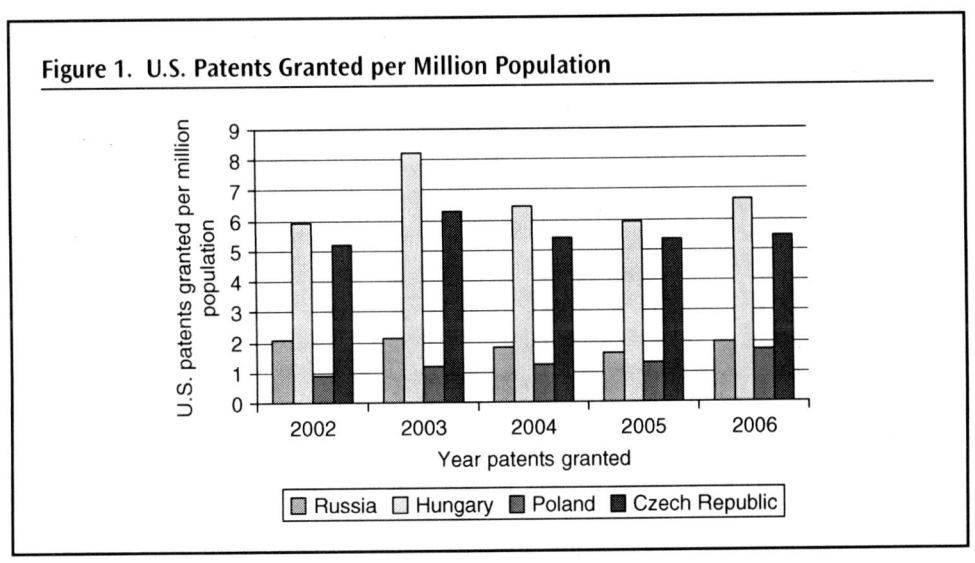

**Figure 1. U.S. Patents Granted per Million Population**

*Sources:* USPTO Cassis CD-ROM, December 2006 version and World Development Indicators, World Bank.

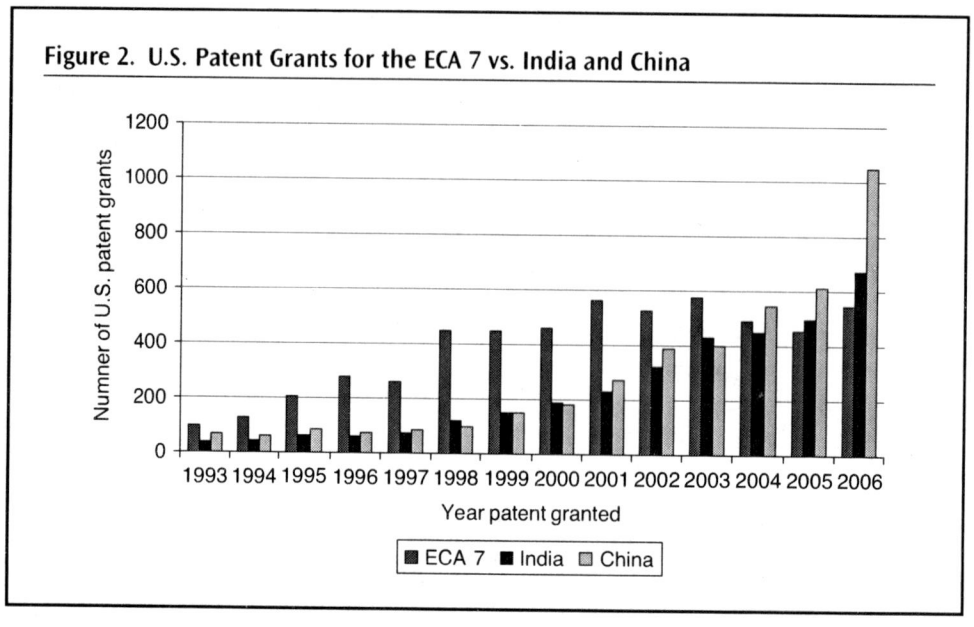

**Figure 2. U.S. Patent Grants for the ECA 7 vs. India and China**

*Notes:* The graph compares counts of patents in which at least one inventor is based in one of seven ECA countries, India, or China. The ECA-7 are Russia, Hungary, Poland, Slovenia, the Czech Republic, Bulgaria, and Ukraine.
*Source:* Authors' calculations based on the U.S. Patent and Trademark Office Cassis CD-ROM, December 2006 version.

From 1993 through the end of 2006, Russia, Hungary, Poland, Slovenia, the Czech Republic, Bulgaria, and Ukraine (ECA-7) obtained 5,489 U.S. patents, whereas India-based inventors obtained only 3,331, and China-based inventors obtained 4,063. The performance of the ECA-7 countries has been much better on a per-capita basis. However, as seen in Figure 2, ECA-7 patenting in the United States has not grown significantly in recent years, while India and China have surged ahead. We also find statistical evidence in many ECA countries of the relative isolation of the R&D community from international technological trends: indigenous patents generally make fewer citations to the existing state of the art than comparable patents filed in other parts of the world, and they cite inventions that have a lesser impact in terms of citation patterns. Moreover, the number of indigenous patents in the ECA region is low relative to the level of R&D investment.[2]

On the positive side, we observe that international R&D collaboration has allowed the ECA region to partially sidestep the handicap of its own low R&D productivity, a problem rooted in the region's insufficient grounding in the recent advances in the state of the art. A large fraction of ECA patents obtained in the EPO are "coinvented" with inventors in Western economies, and Germany plays a particularly important role. A coinvented patent

---

2. While China and India spend 1.23 percent and 0.8 percent, respectively, of their GDP on R&D expenditures see (http://www.financialexpress.com/news/India-lags-China-in-RampD-spending-Sibal/283583/), the ECA 7 countries as a whole spend under 1 percent of GDP on R&D http://europa.eu/rapid/pressReleasesAction.do?reference=STAT/08/34&format=HTML&aged=0&language=EN&guiLanguage=en and ECAKE I report).

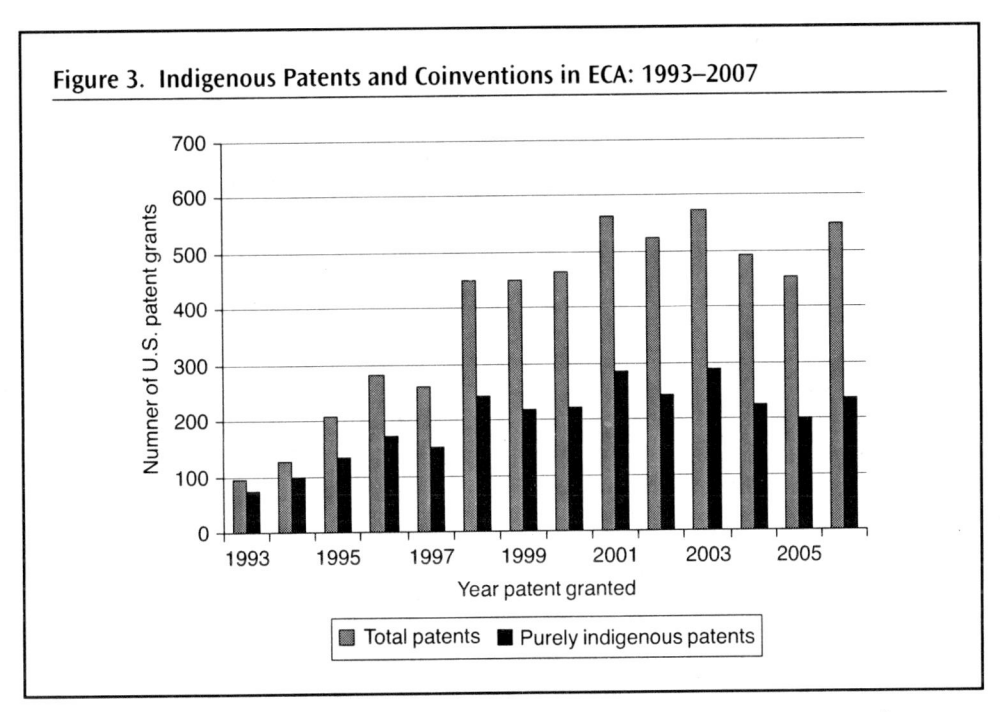

**Figure 3. Indigenous Patents and Coinventions in ECA: 1993–2007**

*Note:* The graph compares counts of patents in which at least one inventor is based in one of seven ECA countries. The ECA-7 are Russia, Hungary, Poland, Slovenia, the Czech Republic, Bulgaria, and Ukraine.
*Source:* Authors' calculations based on the USPTO Cassis CD-ROM, December 2006 version.

is one where at least one named inventor is located in the ECA region, and at least one inventor is located outside the region. Coinventions are quite common (Figure 3). Most regional specialists would not be surprised to see the prominence of German inventors in these coinventions.

## Analysis of Patents and Citations—Policy Implications

The findings reaffirm the need for continued efforts to reform ECA R&D systems and complete the transition from the socialist-era science and technology architecture to a system modeled on global best practices that is more internationally integrated and market driven. We find statistical evidence of the relative isolation of ECA R&D in the citation patterns, as illustrated by the relatively small number of citations these patents receive in patents that are subsequently granted.

Foreign firms appear to be making a significant contribution to ECA-region inventive activity. These firms' local R&D operations, and their sponsorship of local inventors, generate a large fraction of the total patents emerging from ECA countries. This process of international coinvention not only contributes to the quantity of ECA patents, but also raises the quality of ECA inventive efforts. Whereas indigenous ECA patents lag behind other regions in terms of the degree to which they build on prior inventions and extend it, the ECA patents created through multinational sponsorship are better connected to global R&D trends and generally represent inventions of higher quality.

Science and innovation policy in the region should encourage ECA countries to promote international collaboration and should support a greater role for the private sector in knowledge generation. We argue that governments should encourage foreign R&D investment and international R&D collaboration. However, as argued in the first study in this series (ECAKE I), measures to support R&D are ineffective when the key prerequisites—human capital and investment climate—are insufficient and less competitive. This is discussed below.

## The Links of Knowledge Absorption to Trade and FDI

Openness to foreign trade and investment is critical to the process of technological absorption and diffusion, not only for the competitive pressure it exerts on management and corporate governance, but also for the exposure to global best practice technology and management techniques provided to local firms. In order to investigate whether firms benefit from such openness through exporting and inward FDI, we address two questions: Is there "learning by exporting?" And how does FDI affect absorption?

The extensive recent literature and empirical work reviewed in Chapter 3 suggest that increased imports, especially of advanced intermediate goods, facilitate productivity growth (Coe and Helpman 1995; Keller 1998, 2000, 2002; Eaton and Kortum 2001, 2002; and Schiff and others 2002). However, the benefits of exporting are not as clear-cut.[3] Studies by Bernard and Jensen (1995), Clerides and others (1998), and Roberts and Tybout (1997), suggest that there is a self-selection bias among exporting firms, where the more productive firms tend to become exporters, and therefore they do not find conclusive evidence of "learning by exporting." In Chapter 3, however, we explore this phenomenon using econometric analyses and take a different methodological tack than the prior studies.

We conclude that closely linked to trade and openness, by examining the academic literature we find that increased FDI flows, especially participation in producer-driven supply chains, increases the absorption of technology. Finally, there is evidence that increased FDI in business services (such as finance, telecommunications, and transportation services) facilitates increased productivity in the host economy, especially in the sectors that use the services (Fernandes 2007; Arnold, Mattoo, and Javorcik 2007). While tariff barriers on trade are not high in many ECA countries, there remains a large, unfinished "behind the border" structural and institutional policy agenda that would likely increase FDI and trade, and the related technology absorption.

From Figure 4 we see that ECA compares poorly with innovation comparators, according to the World Bank's World Trade Indicators database. The Trade Restrictiveness Index (TRI) captures the barriers to trade, with a higher TRI implying a greater degree of restrictiveness. While the TRI averages between 4 and 5 in several ECA countries, there are negative outliers such as Romania and Russia. Their TRIs are over three times higher than

---

3. While Blalock and Gertler (2004), and Van Biesebroeck (2005), find enhanced productivity enjoyed by developing country firms that export to more technologically advanced countries, this finding is not generally supported in the literature.

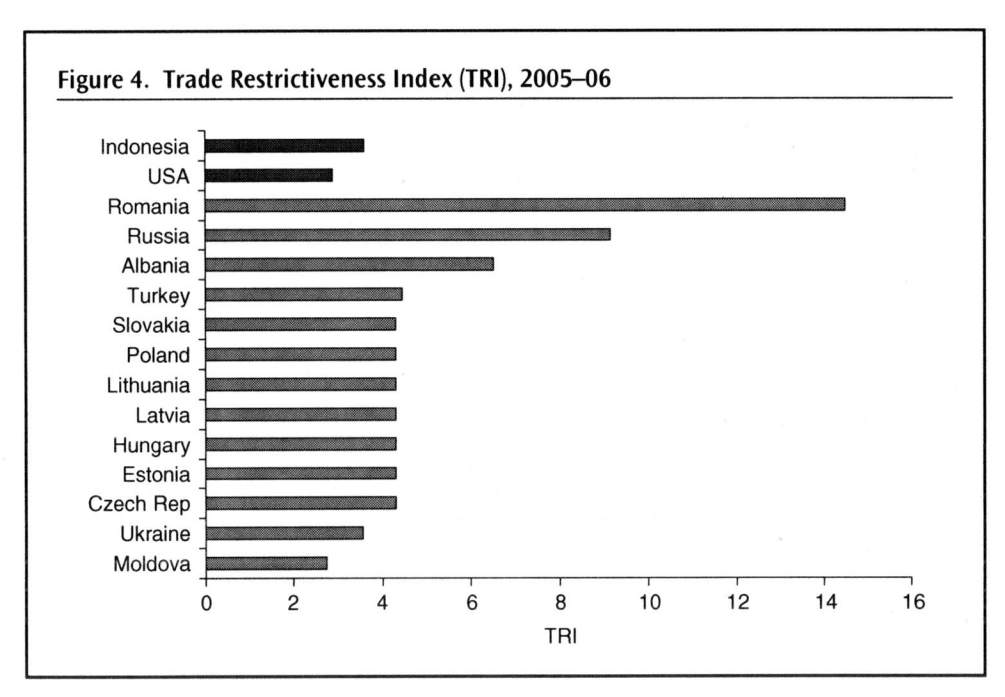

**Figure 4. Trade Restrictiveness Index (TRI), 2005–06**

| Country | TRI |
|---|---|
| Indonesia | ~3.5 |
| USA | ~3 |
| Romania | ~14.5 |
| Russia | ~9 |
| Albania | ~6.5 |
| Turkey | ~4.5 |
| Slovakia | ~4 |
| Poland | ~4 |
| Lithuania | ~4 |
| Latvia | ~4 |
| Hungary | ~4 |
| Estonia | ~4 |
| Czech Rep | ~4 |
| Ukraine | ~3.5 |
| Moldova | ~3 |

*Source:* World Trade Indicators database, 2007.

newly industrializing Indonesia and the TRI of the United States. In terms of FDI inflows (see Figure 5), there is a visible intraregional disparity among ECA countries. The 10 EU member countries from Central and Eastern Europe (EU10) and Southeast Europe (SEE), such as Estonia and Bulgaria, are attractive for FDI. Resource-rich countries such as Azerbaijan, and middle-income countries including Ukraine, are also attractive FDI destinations. However, countries like Belarus and Macedonia have been unable to attract much FDI. Uzbekistan had zero FDI inflows in 2005. The econometric analysis in Chapter 3 aims to look at these trade-related factors at the firm level to analyze their impact on various measures of technology absorption.

## Econometric Analysis

The Business Environment and Enterprise Performance Survey (BEEPS) datasets are cross-sectional surveys, and the most recent one, conducted in 2005, is the most complete. In these surveys, firm managers are asked specifically whether their firm recently introduced a new (to the firm) product, upgraded an existing product, acquired a new production technology, signed a new product licensing agreement, or acquired a new quality certification. Potentially, each of these represents a dimension of the kind of technology absorption process we believe is fostered through exposure to international best practices. The contribution of our analysis lies in establishing a direct connection between the exposure to international technical best practices, through trade and supply networks, and specific, discrete processes of technical improvement at the firm level.

The indicator variables mentioned above are more direct measures of technology transfer or technology absorption than changes in productivity. While the measures have

**Figure 5. FDI Inflows as a Percentage of GDP, 2005**

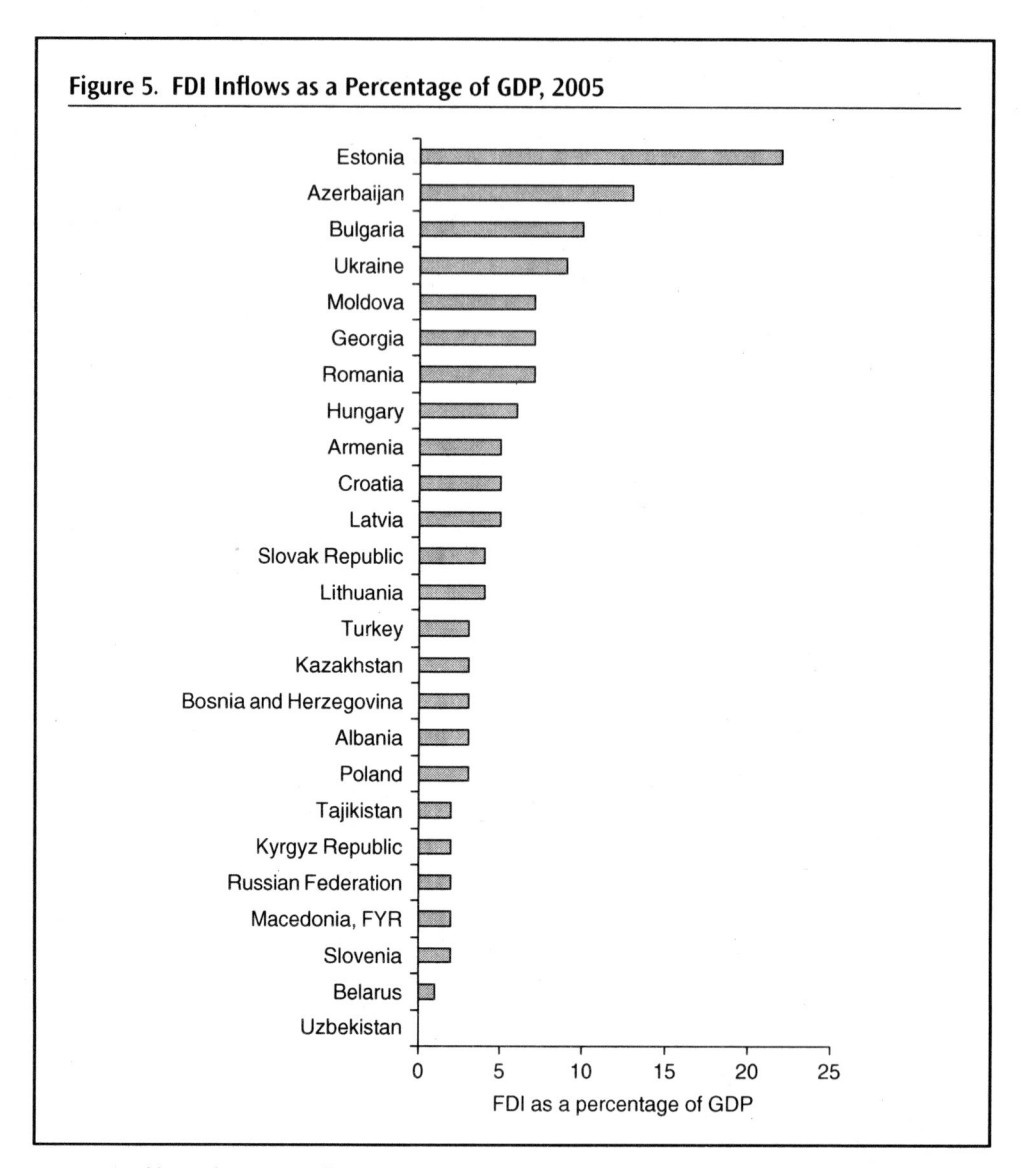

| Country | FDI as a percentage of GDP |
| --- | --- |
| Estonia | ~22 |
| Azerbaijan | ~13 |
| Bulgaria | ~10 |
| Ukraine | ~9 |
| Moldova | ~7 |
| Georgia | ~7 |
| Romania | ~7 |
| Hungary | ~6 |
| Armenia | ~5 |
| Croatia | ~5 |
| Latvia | ~5 |
| Slovak Republic | ~4 |
| Lithuania | ~4 |
| Turkey | ~3 |
| Kazakhstan | ~3 |
| Bosnia and Herzegovina | ~3 |
| Albania | ~3 |
| Poland | ~3 |
| Tajikistan | ~2 |
| Kyrgyz Republic | ~2 |
| Russian Federation | ~2 |
| Macedonia, FYR | ~2 |
| Slovenia | ~2 |
| Belarus | ~1 |
| Uzbekistan | ~0 |

*Source:* World Development Indicators, 2007.

the disadvantage of reflecting the self-assessment of a firm's representative, they have the important advantage of not being obscured by changes in the firm's market environment that are coincident with the absorption of the new technology and may affect productivity measures. All but one of our measures of international connectedness are positively associated with upgrading technology. Notably, the magnitude of the effects of exporting and participating in multinational joint ventures (JVs) is particularly large. Another empirical regularity that merits comment is the robustly positive relationship that exists between measures of human capital at the firm level and technology absorption.

Our enterprise survey analysis and case studies show that firms that sell to multinationals or engage in cooperative activities with multinationals seem more likely to: introduce new (to the firm) products and processes, upgrade existing products and processes, acquire new (to the firm) product or process technologies, and engage in a range of behaviors associated

Table 1. Correlations of "Composite" Measure of Absorption[4] with Firm Characteristics [+] Positive Correlation, [−] Negative Correlation

**Dependent Variable: Upgrade**

| | | | | | | |
|---|---|---|---|---|---|---|
| Exporter dummy | [+]*** | | | | | [+]*** |
| Export as a percentage of sales | | [+]*** | | | | |
| Majority foreign owned | | | [−]* | | | [−]*** |
| Percentage sales to MNCs | | | | [+]*** | | [+]*** |
| Joint venture with MNCs | | | | | [+]*** | [+]*** |
| Size | [+]*** | [+]*** | [+]*** | [+]*** | [+]*** | [+]*** |
| Age | [−]*** | [−]** | [−]** | [−]** | [−]** | [−]** |
| State owned | [−]*** | [−]*** | [−]*** | [−]*** | [−]*** | [−]*** |
| R&D expenditure | [+]*** | [+]*** | [+]*** | [+]*** | [+]*** | [+]*** |
| Internet use | [+]*** | [+]*** | [+]*** | [+]*** | [+]*** | [+]*** |
| Training | [+]*** | [+]*** | [+]*** | [+]*** | [+]*** | [+]*** |
| Skilled workforce | [+]*** | [+]*** | [+]*** | [+]*** | [+]*** | [+]*** |
| University graduates | [+]* | [+]** | [+]** | [+]** | [+]* | |
| Governance index | [+]*** | [+]*** | [+]*** | [+]*** | [+]*** | [+]*** |
| Use of loan | [+]*** | [+]*** | [+]*** | [+]*** | [+]*** | [+]*** |
| Observations | 7901 | 7897 | 7920 | 7754 | 7920 | 7736 |
| R-Squared | 0.26 | 0.25 | 0.25 | 0.26 | 0.28 | 0.29 |
| Country dummies | Yes | Yes | Yes | Yes | Yes | Yes |
| Industry dummies | Yes | Yes | Yes | Yes | Yes | Yes |

\* Significant at 10%, \*\* Significant at 5%, \*\*\*Significant at 1%.
Source: Authors' calculations using BEEPS database.

with increasing technological sophistication. The Serbian case studies in Chapter 4 further suggest that absorption requires tough decisions and large investments, as firms need to spend resources on modifying imported equipment and technologies, and reorganizing production lines and organizational structures. The case studies highlight the important role of foreign investors in knowledge absorption, whether acquired through capital goods imports, exporting, hiring consultants and other knowledge brokers, or from licensing technology.

## Is There "Learning by Exporting?"

The concept of learning by exporting has been seen in the literature as a process by which exporting increases productivity by exposing producers to new technologies, or through upgrading product standards and quality. Exporting is another channel through which firms based in open economies can acquire foreign knowledge about technologies and

---

4. The dependent variable "upgrade" is the sum of firm responses to questions on 1) introduction of a new product or process, 2) upgrading of existing product or process, 3) achievement of new quality certification, and 4) new technology licensing agreement.

products. However, the apparent absence of statistically significant learning-by-exporting effects documented in many contexts has led many economists to question its importance, and has, to some extent, undermined the notion that reforms that favor open trade will increase corporate productivity.

Our study strongly suggests that the conventional methodology employed by virtually all previous researchers to estimate learning-by-exporting effects may be biased against finding a linkage between exports and increased technology absorption, even when such a relationship exists. This bias exists, in part, because the conventional methodology uses changes in TFP as a proxy for technology absorption and diffusion, instead of considering concrete instances of efficiency-enhancing technological investment.

In our panel data, transition to exporting is positively and significantly correlated with increases in measured upgrading of technology. We find this to be true even after we control for firm size, firm age, state ownership, foreign ownership, human capital, and environmental factors affecting the export climate. This is consistent with the hypothesis that exposure to foreign markets fosters learning, and our results suggest that this learning effect is not limited to foreign-owned firms. It is, of course, not inconsistent with the view that firms, as they seek to transition to exporting, will invest in upgrading their technology to make themselves more competitive in foreign markets. In other words, upgrading of technology could also increase exports—but, to the extent that the upgrade was initially motivated by the desire to compete in a foreign market, it still reinforces the policy implications we stress. These findings are supported by studies in ECA, most recently by de Loecker (2007), who shows that Slovenian firms learned from exporting.

## *How Does FDI Affect Absorption?*

There is an ongoing debate in the literature about the impact of FDI: Some have suggested that positive technology spillovers from FDI are largely limited to "vertical" FDI transactions, in which there is a direct purchasing relationship between the foreign firm and the local supplier. Using more direct proxies of the firm's relationships with multinationals, we find evidence that vertical FDI promotes learning by local firms. Our case study in Chapter 4 identifies some explicit channels through which learning occurs.

Reduction of the remaining barriers to FDI in ECA could increase FDI and, given the positive relationship between absorption and FDI, facilitate absorption. For example, Russia fares worse than other countries in the region, attracting one of the lowest per-capita levels of FDI inflows. World Bank research has pointed to key shortcomings in the Russian business environment. Many of these are a function of government policies that limit FDI inflows and foreign firm operations, especially in the services sector. Another ECA country where reform is needed is Kazakhstan, which has done more to lower its tariffs on goods than it has to liberalize its barriers to FDI in the service sector.

Technological advances and economic liberalization, while creating new markets and opportunities, have also led to the need for an efficient and timely global production and logistics network. High logistics costs are a barrier to trade and foreign direct investment. In view of the importance of exports and FDI to learning and knowledge absorption in the private sector, the "behind the border" reform agenda in ECA continues to play a crucial role in enlarging the available channels for technology absorption from the rest of the world. The reform agenda in ECA includes reduction of entry barriers, reforms of the regulatory regimes for business services, trade facilitation, (for example, customs procedures, addressing governance problems for transportation services), and reduction of nontariff and technical barriers to trade.

## How Does FDI Stimulate Knowledge Absorption? A Case Study of Serbian Enterprises

The objective of the case study is to complement the findings from the econometric analysis of the BEEPS surveys summarized above, which provides evidence in favor of the view that multinationals contribute to *indigenous* technological improvement. A case study approach can provide only a richer perspective on the causality between FDI and absorption. Focusing on the dimensions of product mix, production technology, management, and skills, the study sheds light on the firm-level absorption process following FDI acquisition in the context of a transition and post-conflict country, Serbia. The results illustrate the critical role played by the foreign strategic investor in helping a company cope with the challenges of absorbing knowledge embedded in capital goods, from exports, from "knowledge brokers," and codified in intellectual property.

Eight large companies operating in the metal processing, household chemicals, pharmaceuticals, and cement industries were included in the case study. In brief, the guiding selection criteria were: industries, as well as company characteristics, (especially firm size) helped for the *comparability* of results; company characteristics and the type of acquisition (especially the type of buyer) provided some *controls to test counterfactuals;* and companies were *privatized early on* to ensure availability of archival information (due diligence, post-acquisition monitoring) and for the effects of restructuring and investment decisions to be discernible in financial data. The chapter presents the main results about absorption via acquisition FDI in different industries, comparing the pairs of companies as far as it is possible.

The case study suggests that the speed in reestablishing a presence in foreign markets that is comparable to or exceeds pre-sanctions exports differs between enterprises privatized to a foreign strategic investor and those sold to a local investor. In general, companies sold to domestic investors were not able to increase exports in a significant way, while comparable firms receiving FDI did much better. Moreover, the more significant changes in product mix and manufacturing occurred in companies bought by foreign investors.

### Table 2. Comparing the Effects of Privatization on FDI versus Domestically-owned Firms

| Sector | Number of Per Employees | | Total Operating Income Per Employee | | Salary Per Employee | | Value Added | |
|---|---|---|---|---|---|---|---|---|
| | Foreign | Local | Foreign | Local | Foreign | Local | Foreign | Local |
| Aluminium processing | −32% | −23% | 308% | 152% | 170% | 64% | 31% | −15% |
| Household chemicals | −27% | −22% | 188% | 24% | 130% | 29% | 53% | 0% |
| Pharmaceuticals | −31% | −11% | 77% | 37% | 99% | 120% | 11% | 9% |
| Cement | −45% | | 199% | | 141% | | | |

*Source:* Authors' calculations.

We found that, among the recipients of foreign investment, all but one company had replaced their top management. New directors were brought in from the multinational enterprise (MNE), the domestic investor's holding, from rival companies, or promoted from within. In companies acquired by foreign investors, the comparative advantage for R&D lies in the adaptation of products and machinery to local conditions, rather than in innovation. For example, advanced formulas or product designs are transferred from the MNE and adapted locally so that the products can be manufactured efficiently in the acquired plant. Local investors have generally cut back on R&D, and instead focus on quality control and marketing.

An analysis of financial results pre- and post-acquisition indicates that productivity, as measured by value added per employee, increased more in the foreign-owned firms than in the locally-owned companies. In Chapter 4, we show that physical productivity (units of production per employee) diverges significantly in the period following the acquisition, from being roughly similar, to being around four times greater in one company bought by a foreign investor. The divergence is even greater when one examines individual product lines, supporting the hypothesis that foreign investors have incentives to focus on core product lines and close others, even if adjustment costs are greater. We find that the decline in post-privatization employment is, as expected, larger in the foreign-owned companies than in locally owned companies, and the post-privatization increase in income per employee is higher in the foreign-owned companies.

The case study targets FDI based on acquisition of existing assets from the government (privatization), or from private owners, rather than "Greenfield"[5] FDI. The importance of this distinction is that countries poor in natural resources (like Serbia), and whose investment climate is still uncompetitive, have a hard time attracting Greenfield FDI. Therefore, attracting FDI by selling existing assets, whether privately held or state owned, is a more realistic option. Importantly, the experiences of the companies we examine show that FDI via acquisition is often a first step for making Greenfield investments.

Our emphasis on the outcomes precipitated by the acquisition of productive assets in Eastern Europe is highly relevant today because of the legacy of mass privatization, which resulted in insider control that prevents openness to change and hampers absorption and innovation. Experience in ECA shows that concentrated ownership, in the form of an oligopolistic market structure or a monopoly, can hamper productivity growth and pose challenges for corporate governance. The premise behind the privatization programs in the 1990s in early ECA reformers, and in the early 2000s in Serbia, was that local expertise would be insufficient to restructure the large companies. These were viable in a closed economy due to the lack of effective competition, but proved to be a problem for global competitiveness. FDI was seen, at least in Central and Eastern Europe and Southeastern Europe, as the only way to reintegrate the companies into European markets.

We know from the literature that risk–taking incentives—necessary ingredients for innovation—depend on good corporate governance. With state or dispersed ownership, managers of enterprises have little incentive to take risks. The Serbian case studies provide insights into the relevance of post-transition corporate governance for the willingness of

---

5. Greenfield investment is *de novo* investment in a previously undeveloped site, where no facilities or plant existed.

managers to innovate and absorb; i.e., to take risks and carry out needed reorganization. The dispersed ownership resulting from mass privatization (or from the pre-reform 1997 law *on ownership transformation* in Serbia) have proved particularly problematic in post-conflict countries plagued by ethnic and social divisions, such as Bosnia and Herzegovina, Moldova, Armenia, FYR Macedonia, and Tajikistan. In such circumstances, a strategic owner, whether foreign or local, is a *sine qua non* condition for improving corporate governance, and consequently, for technology absorption.

The spillovers of FDI to other firms, or to the whole economy, whether by diffusion of technological and other knowledge, or by sharpening the incentives for domestic firms to upgrade their productivity, play a central role in the economic literature. However, such spillovers are secondary in the early days of transition. The critical role of Brownfield FDI is to restructure the acquired companies as expeditiously as possible. Early in the transition process, indigenous firms need the help of foreign investors to acquire new technical competencies. In such cases, the beneficial role of FDI is seen in the company in which the foreign firm has invested. This occurs even when rival enterprises or the economy in general are not yet ready for the conventionally defined spillovers. While the econometric analysis looks at technological absorption patterns in countries long after transition began, the case studies analyze firms soon after the transition process, lending interesting insights.

## Postscript on Challenges Ahead for ECA

This study provides persuasive new evidence of the importance of trade openness, FDI, human capital, R&D, and knowledge flows for innovation and absorption in ECA. The countries in the region differ in their remaining reform agendas, both in general terms and in these areas in particular, as well as in the relevant resource endowments needed to effect the transition toward a more dynamic and globally competitive knowledge economy. The main lessons from the study would suggest that additional efforts toward the following objectives would contribute to knowledge absorption in ECA:

- Intensified international R&D collaboration and foreign R&D investment to enhance the integration of ECA in the global R&D community;
- Progress on the unfinished "behind the border" trade reform agenda to increase the openness of several ECA countries to global trade networks; and
- Further opening to FDI to play an important role in encouraging knowledge absorption, notwithstanding the opposition to FDI in some ECA countries.

The rapid technological catch-up experienced in the more reformed economies is indicative of the fact that within-firm productivity improvements, rather than reallocation of resources, are increasing in importance as a source of growth in ECA. The countries that are now member states of the European Union increasingly participate in FDI-driven intraindustry trade, which is conducive to technology absorption. But many of the Commonwealth of Independent States (CIS) and SEE countries still need to restructure enterprises, ease the entry and exit of firms, improve access to credit, and accelerate the "behind the border" reforms to benefit from trade openness, and thus from technology absorption.

# CHAPTER 1

# Introduction

*By Itzhak Goldberg and Smita Kuriakose*

A key driver of economic growth and industrial development is a country's absorptive capacity, or ability to tap into the world technology pool. Trade flows, foreign direct investment (FDI), and labor mobility and training are among the best conduits for knowledge absorption. But such transfers are not automatic. They also require a favorable investment climate, an educated workforce, and, not infrequently, some research and development (R&D) on the part of the absorbing country. This study provides an analysis of the extent of knowledge and technology absorption by firms in Europe and Central Asia (ECA), based on the statistical analyses of various data sources, including World Bank Enterprise Surveys and the patent databases maintained by the United States and European patent offices, as well as original case studies.

## Definition of Innovation and Knowledge Absorption

As a point of departure, we refer to the ideas in Cohen and Levinthal (1989, 1990), and Griffith, Redding, and Van Reenen (2004), which shape the following definition of absorptive capacity at the firm level:

> A firm's capacity to assess the value of external knowledge and technology, and to make the necessary investments and organizational changes to absorb and apply this in its productive activities.

Absorption is therefore a costly learning activity that a firm can employ to integrate and commercialize knowledge and technology that is new to the firm, but not new to the world. The development of new-to-the-world knowledge can be considered innovation.

In other words, innovation shifts a notional technological frontier outward, while absorption moves a firm closer to the frontier. Examples of absorption include: adopting new products and manufacturing processes developed elsewhere, upgrading old products and processes, licensing technology, improving organizational efficiency, achieving quality certification, etc.

This study is a follow up to *Public Financial Support of Commercial Innovation* (ECAKE I), which focused on innovation. As indicated in ECAKE I, this followup study provides a more detailed analysis of the complementarities between innovation and absorption, with a focus on absorption. As will be shown in the following chapters, based on the literature (Cohen and Levinthal) and from our empirical findings, R&D, which is a key input into innovation, is also an input into absorption. Indeed, there are important complementarities between innovation and absorptive capacity. Innovation promotes absorptive capacity because the generation of human capital and new ideas, and the associated knowledge spillover effects, help build absorptive capacity. Conversely, the absorption of cutting-edge technology inspires new ideas and innovations.

While recognizing the general importance of policies that would improve the absorptive capacity of firms in ECA, this study's point of departure is that policies need to be designed according to the level of development of countries, and specifically, depending on whether these policies are meant for advanced reformers or for less advanced ones. Consistent with this view, the study, *Productivity Growth, Job Creation, and Demographic Change in Eastern Europe and the Former Soviet Union* (World Bank 2007b), points out that late reformers will need to continue focusing on facilitating the restructuring and the entry and exit of firms, improving access to credit, and accelerating the "behind the border" reform agenda, in order to benefit from trade openness. The emphasis is somewhat different for advanced reformers, which need to focus on how knowledge and private sector-led innovation and R&D can be harnessed to sustain productivity growth within the firm, which is the focus of this study.

## Diffusion of Knowledge in Support of Productivity Growth: Literature Review

Ever since the path-breaking research of Robert Solow (1956), economists have known that secular growth is closely connected to technological change, in addition to factor accumulation. A vast array of empirical research during the last half century has shown conclusively that a substantial share of growth in income per capita is associated with the growth of total factor productivity (TFP). Attaching the label of technological change to Solow's famous "residual" (TFP growth), begs the question of what technological change contains exactly, and perhaps, more importantly, how it evolves over time, as well as the nature of the economic forces that determine its course and pace.

One of the challenging aspects for economists who first considered these matters was that the growth of TFP appeared as an impenetrable "black box" that operated outside the realm of economic forces. A long and very fruitful research agenda pioneered by Arrow, Abramovitz, Griliches, Jorgenson, Denison, Rosenberg, and others, sought to open this black box, and to explain the economic drivers of technological change. With the development of the endogenous growth theory in the late 1980s and 1990s (Romer 1986, 1990;

Lucas 1988; Grossman and Helpman 1991; Aghion and Howitt 1995), the economic profession as a whole came to accept the view that innovation, knowledge spillovers, and R&D were indeed key factors driving self-sustained, long-term economic growth. It also became widely accepted that these activities were generated from within the economic system, which was responding to economic incentives.[6]

It is fair to say that the early contributions and endogenous growth literature were, to a large extent, focused on *new-to-the-world* activities. Meanwhile, a separate stream of literature examined the processes of diffusion and absorption of new technologies. This literature owes much to the pioneering contribution by Griliches (1957) about the timing and location of technology adoption, and his paper was based on empirical research on the spread of new hybrid corn varieties in U.S. agricultural regions.

The two approaches have recently converged in the Schumpeterian growth theory, which has investigated the factors that determine the absorption of technology by firms that are behind the "technology frontier," in contrast to the incentives behind radical innovations. An example is Aghion and other (2002), which examines the interplay between competition and innovation, and its impact on growth. For firms that are at similar technological levels ("neck-and-neck" industries), the Aghion model shows that competitive pressures will stimulate the incentives for firms to invest in R&D, with the aim of increasing their competitiveness; and the same reasoning would apply when managers face hard budget constraints. In contrast, when followers are far from the technology frontier, competitive pressures have a weak effect on the R&D incentives of technological leaders, who are so far ahead of the pack, and have no motivation to increase innovation activities. The prediction that absorption incentives are weak in "unleveled" industries—those in which followers are using obsolete machinery and equipment, unskilled workers, etc.—can be argued as applying to certain industries in transition countries, where foreign entrants are much more technologically advanced. Keller (2002) discusses how international technology diffusion relates to other factors affecting economic growth in open economies, and the importance of specific channels of diffusion, particularly trade and foreign direct investment.

In the innovation literature, it is well established that market failures restrict private investment in innovation, justifying public intervention. Several arguments can also be made in favor of government support for technology absorption. For example, Hausman and Rodrik (2003) reason that countries lagging behind the technology frontier will underinvest in technology absorption because of the shortcomings of their intellectual property regime. Unlike innovating companies, which are protected by patents, entrepreneurs who invest funds to discover which technology to adapt to a developing country context do not normally get any protection for the markets they open, no matter how high the associated social return may be.

Underlying the above discussion is the idea that uncertainty and imperfect information serve to obscure and impair the returns from technology absorption. Because of this shortage of information, firms attach such value to the second face of R&D (to paraphrase Cohen and Levinthal, 1989): this relates to the role of R&D in identifying, collecting, and

---

6. Work on education and technological change by Nelson and Phelps (1966) mentioned that technological progress was the key to growth and highlighted the difference (for growth) between human capital stock and accumulation. However, it was only in the late 1980s that those views were widely shared.

analyzing information regarding the feasibility of technological absorption, as well as support for its implementation. There is evidence from Griffith, Redding, and Van Reenen (2004)—based on a survey of industries across 12 OECD countries—which shows that R&D enhances technology transfer by improving the ability of firms to learn about advances on the technology frontier. Thus, R&D is important both in this catch-up process, as well as in directly stimulating innovation. It is important to mention standards for technical interoperability, quality control, etc., as they have been instrumental in enabling cross-country diffusion of technology.

*Global Economic Prospects 2008* (World Bank 2008) argues that geography and history appear to be important determinants of technological achievements, and countries in ECA show significantly higher levels of technological achievements than do other countries at similar income levels. It shows that the technology gap between middle-income and high-income countries has narrowed over the past 10 years, and that evidence of catch-up is particularly strong in the Czech Republic, Hungary, and Poland, where the level of technological achievement increased by more than 70 percent during the 1990s. *Global Economic Prospects 2008* develops an aggregate measure of technological achievement using a statistical technique (principal components analysis) that combines some 20 separate indicators of technological achievement along three dimensions—scientific innovation and invention; penetration of older technologies; and penetration of newer technologies—plus an additional dimension, the extent to which countries are exposed to external technologies. Distribution of overall technological achievement across countries, and the change over the past decade, are examined to evaluate both the speed with which technological achievement in countries is advancing, and the dimension along which the change is occurring.

## Conceptual Framework: Economic Conditions and Capacities for Knowledge Absorption

In accordance with the literature reviewed above, the conceptual framework that molds our analysis assumes that innovation and the resultant absorption of productive knowledge and technology are central forces behind economic growth; these forces play just as important a role as the accumulation of human capital and physical capital. Key inputs and economic conditions enable and motivate innovation and absorption activities, and these are central to our endogenous production function approach (see Figure 1.1). Human capital availability and a favorable investment climate are two important prerequisites for firms to be able to innovate because it would be very costly to carry out and reap the benefits from innovation and absorption if these necessary conditions were not in place. Our analysis will use different methods to explore how such factors as human capital and investment climate variables, among others, shape absorption decisions at the firm level.

Trade, FDI, R&D, and knowledge flows are "channels of absorption," by which we mean that they are central conduits for cutting-edge innovations and good practices that are to be diffused to other countries, and then to ultimately percolate into the private sector. Figure 1.1 describes the important channels of absorption at the country, as well as the firm, level. While in and out migrations of the workforce and brain circulation are important channels of innovation and absorption, they are beyond the scope of the analysis in this study. The channels, highlighted in the middle column of Figure 1.1, will constitute

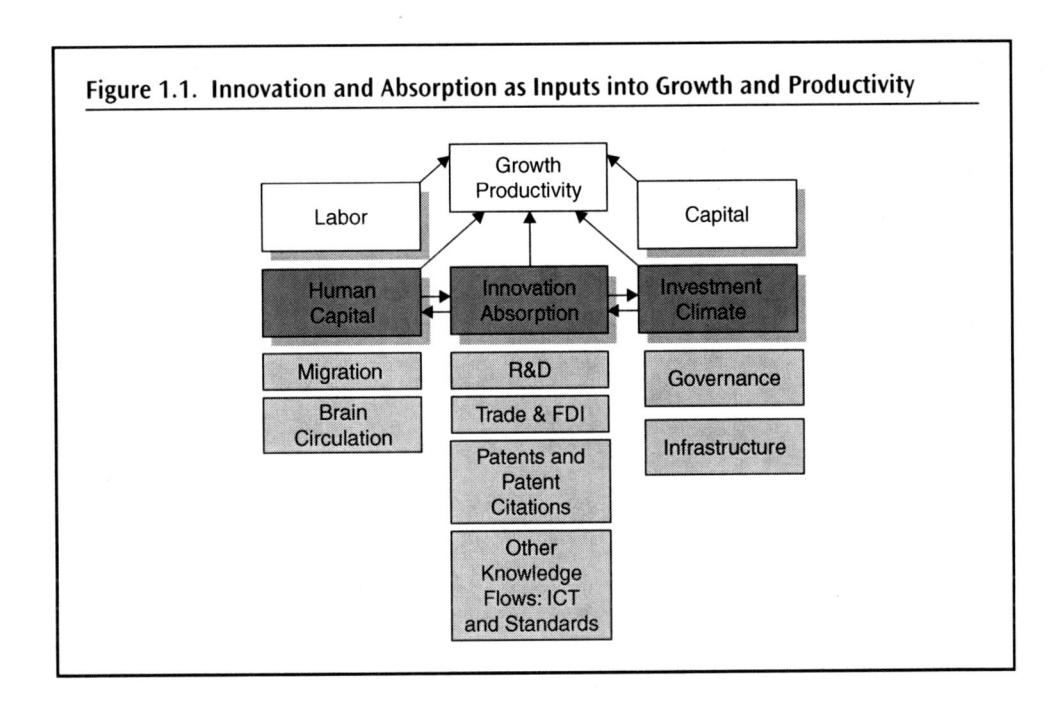

**Figure 1.1. Innovation and Absorption as Inputs into Growth and Productivity**

```
                          Growth
                        Productivity
         Labor                              Capital

       Human          Innovation        Investment
       Capital        Absorption          Climate

      Migration          R&D            Governance

       Brain          Trade & FDI
     Circulation                        Infrastructure
                    Patents and
                       Patent
                      Citations

                       Other
                     Knowledge
                    Flows: ICT
                    and Standards
```

the central focus of the study. Our analysis will primarily investigate how the presence of these channels (the fact that a given firm has received FDI, or invests a certain amount in R&D), affects absorption outcomes. As a secondary question, we will examine the firm-, industry- and country-level conditions that induce firms to activate these channels as part of their overall growth and sales strategies.

Just as the prerequisite conditions and capacities are impacted by governmental policies, so too can properly designed economic policies influence the channels of absorption and the specific decisions for firms to absorb technology. Technology absorption needs a stable and conducive policy framework, and a firm's ability to absorb this technology and knowledge depends on its inherent characteristics, such as resource base, R&D expenditure, and worker skills. Access to knowledge and technology is increasingly linked to FDI and trade. However, extracting benefits from these channels requires dynamic local firms and institutions. Firms' absorptive capacity is thus determined by (i) conditions internal to the firm, such as the presence of foreign investors, foreign trade, and skill endowments; and (ii) conditions external to the firm, such as the costs and incentive structure determined by the wider environment (notably the regulatory framework; openness to knowledge flows, trade and FDI policies; the quality, availability, and cost of infrastructure services; and the ease of access to finance). These aspects of the investment climate are being studied in ECA countries via Investment Climate Assessments (ICA) conducted by the World Bank; thus, in addition to our regional analysis, we rely on the ICAs to provide in-depth national assessments.

CHAPTER 2

# Patents as Indicators of Technological Activity in the ECA Region

*By Lee Branstetter*

## Patent Data Provide a View of the Knowledge Absorption Process

Our study takes as its primary focus the absorption of technology by ECA enterprises, rather than the creation of fundamental new technology. The reason for this focus is clear. Many ECA firms and industries lag well behind the global technological frontier, and relatively few ECA firms or industries are so technologically sophisticated that they could expect to play a leading role in the advancement of that frontier, at least in the near term. Given the level of development of the region's economies, it is almost certainly more important for managerial effort and public policy to be more focused on convergence with the global frontier than on support of indigenous attempts at fundamental innovation. Nevertheless, as we have stressed throughout the report, technological absorption is not a passive process that proceeds automatically with no effort on the part of the absorbing firm. Rather, it is often the case that extensive, active efforts are required to take technology pioneered outside the region and adapt it—in large and small ways—to the economic circumstances of ECA countries. Once again, we invoke the Cohen-Levinthal notion of absorptive capacity: a firm must be engaged in an active process of learning about technologies in order to effectively absorb advances in these technologies by other firms. In some instances, the processes of modification and adaptation lead to innovations, often incremental in nature, dramatically increasing the value of the underlying technology in an ECA context.

There are industries, firms, and regions within the ECA countries where the process of technology absorption has proceeded far enough that this kind of incremental innovation is taking place on a reasonably large scale. To a greater extent than is commonly realized, the major patent systems often grant patents that protect even relatively incremental innovations—both in terms of products and processes. These patents, and the wealth of detailed information they contain, offer a useful window into the ECA technological absorption process that we explore in this chapter.

Almost by definition, successful patenting requires that a firm understand the existing state-of-the-art technology well enough to improve upon it, albeit perhaps in incremental ways. Simply by observing the firms, regions, and industries in which ECA inventors are most active, we can obtain objective, quantitative information on where the technological absorption process is most advanced. We can also observe how the locus of absorption and invention has shifted over time. The significant changes in ECA country patent regimes mean that the patent statistics of the region's countries themselves are unlikely to offer a consistent measure of inventive output over the course of the reform process. For that reason, we rely on data generated by ECA inventors seeking patent protection in the world's two largest patent jurisdictions: the European Patent Office (EPO) and the U.S. Patent and Trademark Office (USPTO). Both organizations provide large quantities of data on ECA-region inventors, obtained through a system whose essential features have remained stable throughout the transition period.

As we have stressed throughout this study, international engagement and connectedness can foster technology absorption and learning. As we will see in our review of patent data, a significant number of ECA-generated patents in Europe and the United States appear to arise out of various forms of R&D collaboration or cooperation with multinational enterprises. A significant fraction of European patents with at least one ECA-based inventor also involve parties from other regions, including Europe's advanced economies. As we will see, Germany plays a particularly prominent role here. Examination of ECA-generated patents in the United States reveals that a significant number of ECA-invented U.S. patents are assigned to American or other foreign multinational firms, a legal arrangement often signifying the existence of an R&D contract or the operation of an R&D subsidiary within the ECA region. These "multinational" patents are generally of higher quality than "indigenous" patents.

Data generated by the U.S. Patent and Trademark Office provide a further unique window into the technology absorption process. Under U.S. patent law, all patent applicants are required to disclose knowledge of the "relevant prior art" on which they are built. These disclosures take the form of citations to earlier inventions and other technical advances that are often the technological antecedents of the invention for which the applicant is seeking patent protection. A large literature has utilized the citations in U.S. patent documents as direct indicators of knowledge spillovers. A recent World Bank Working Paper (Brahmbhatt and Hu 2007) uses patenting in the United States as an index to assess East Asian prowess in generating innovations that advance the global frontier of knowledge. The study looks at patents as an output of innovative activity (number of patents), as well as a measure of patent quality based on the Henderson, Jaffe, and Trajtenberg (1999) approach for measuring the quality of patents, by constructing indexes of patent generality and patent originality that are based on analyses of patent citations.

In other words, the citations explicitly trace out the pathways by which inventors absorbed useful knowledge from prior inventions, and used this knowledge to create new inventions. While the more recent literature has clarified important limitations in the degree to which patent citations reflect pure knowledge spillovers, it has also validated the usefulness of citations as a window into the knowledge absorption process. European patents also contain citations to prior inventions, but because European patent law does not require disclosure by the applicant, the vast majority of European patent citations are added *ex post* by patent examiners, and may or may not reflect inventions that were a source of inspiration to—or even known by—the actual inventor. Interestingly, detailed examination of the citation patterns in indigenous ECA patents reveals significant contrasts between ECA patents and those of other developing regions. Indigenous ECA patents tend to be systematically less well connected to high-quality prior research than do patents from the more dynamic parts of the developing and developed world. This appears to validate the widely held view that ECA inventors, while highly skilled and well educated, are insufficiently connected to centers of technological excellence outside the region to reach their potential levels of research productivity.

## General Trends in ECA Patenting

First, it is clear that the transition process of the early to mid-1990s disrupted inventive activity in the short run. Both U.S. and European patents reveal a striking downturn in inventive output during these years. By the mid-1990s, however, measures of inventive output were trending upward, and this generally positive trend has been maintained up to the most recent years for which data are available. Graphical evidence of this is provided from two very different samples of the patent data.

Figure 2.1 gives a long-run perspective on ECA patenting by tracking all patent applications (including those later withdrawn or denied) for all EPO patents with at least one inventor based in the ECA region. The figure shows a substantial increase in patent activity,

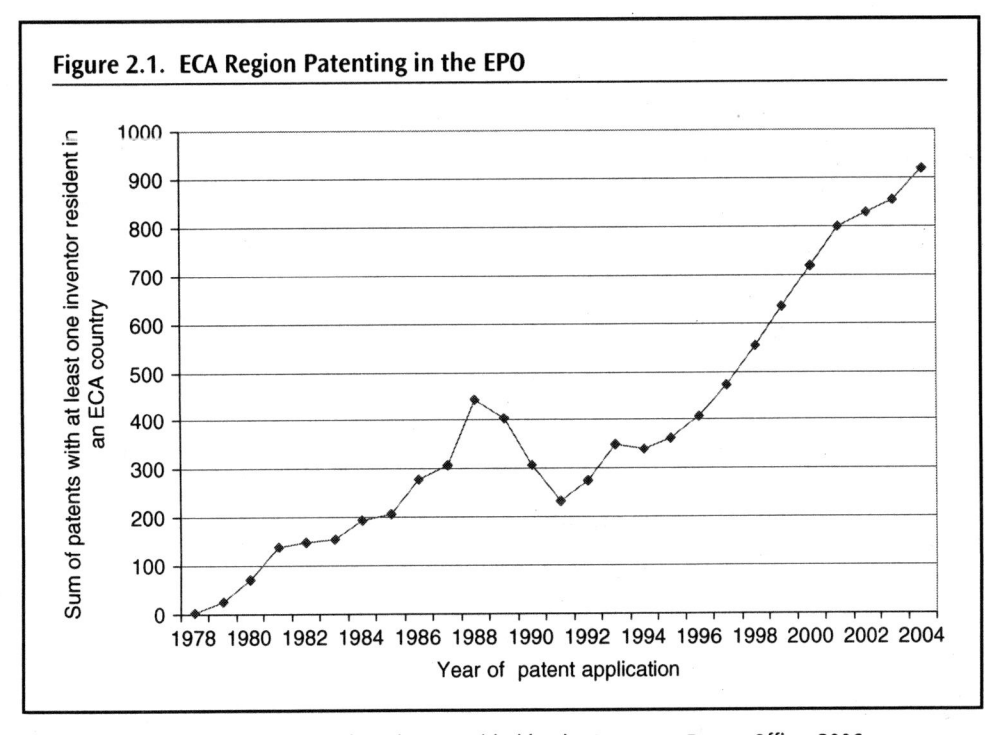

**Figure 2.1. ECA Region Patenting in the EPO**

*Source:* Authors' calculations based on data provided by the European Patent Office, 2006.

peaking in the late 1980s, followed by an equally impressive decline. While levels of patenting outside the region continued to climb, inventive activity within the region did not reattain its late 1980s level until the late 1990s. Since then, there has been robust growth. This pattern for the region as a whole is replicated in similar charts one could produce for individual ECA countries, such as Hungary or Russia/the former Soviet Union. One sees a similar pattern in data on ECA-generated patents registered with the U.S. Patent and Trademark Office. It is clear that the transition process massively disrupted inventive activity.

Second, measures of inventive activity suggest a disproportionate concentration of that activity in the relatively more advanced ECA economies. Perhaps not surprisingly, Hungary and the Czech Republic are clear standouts in terms of inventive performance. Russia is a large patent generator in aggregate terms, but it is less significant than one might expect given its size and Cold War era scientific strength. Figure 2.2 provides a breakdown based on EPO patent applications with at least one ECA inventor, registered between 1992 and 2005.

Third, various forms of international R&D cooperation appear to be quite important in ECA inventive activity. A large fraction of ECA patents taken out in the EPO are coinvented with inventors in Western economies, of which Germany plays a particularly important role. Figure 2.3 provides a breakdown based on EPO patent applications taken out from 1992 through 2005. We designate a coinvention as a patent in which at least one named inventor is located in the ECA region and at least one inventor is located outside the region. Coinventions are quite common. Most regional specialists would not be surprised to see the prominence of German inventors in these coinventions.

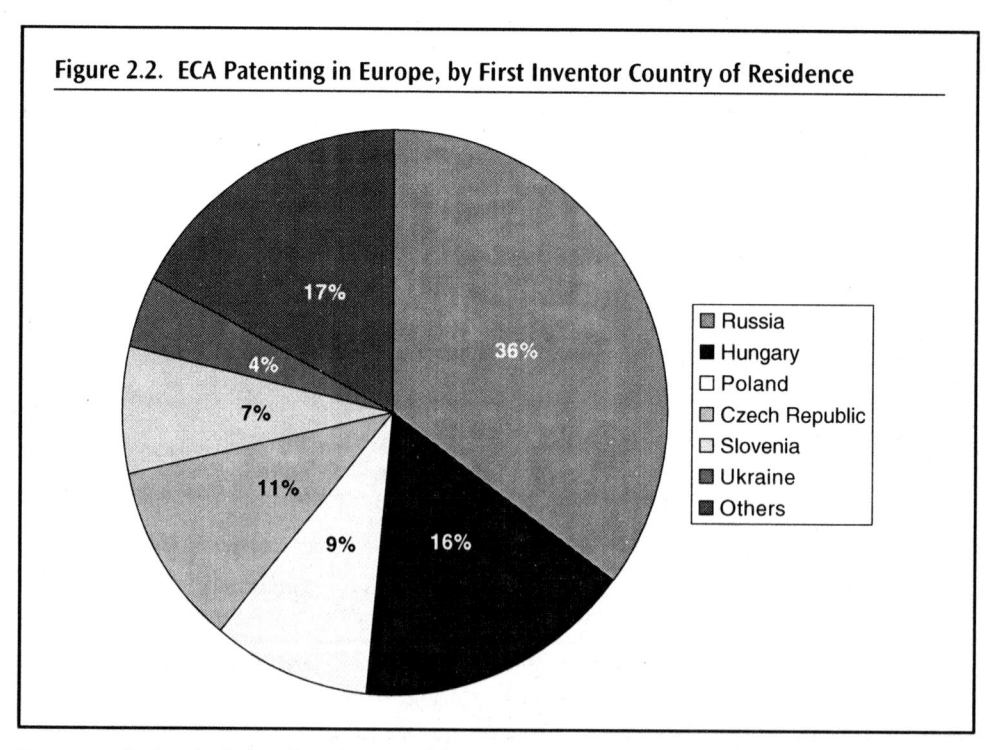

Figure 2.2. ECA Patenting in Europe, by First Inventor Country of Residence

- Russia: 36%
- Hungary: 16%
- Poland: 9%
- Czech Republic: 11%
- Slovenia: 7%
- Ukraine: 4%
- Others: 17%

*Source:* Author's calculations based on data from the European Patent Office, 2006.

**Figure 2.3. ECA Coinvention by Partner Country**

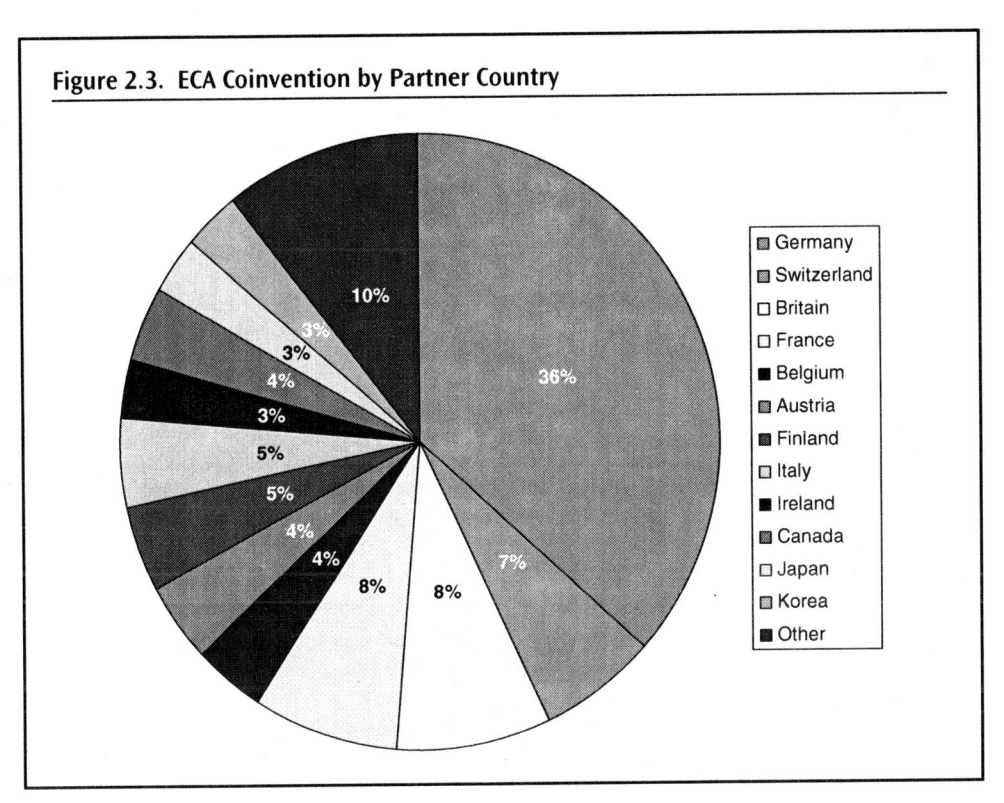

- Germany: 36%
- Switzerland: 7%
- Britain: 8%
- France: 8%
- Belgium: 4%
- Austria: 4%
- Finland: 5%
- Italy: 5%
- Ireland: 3%
- Canada: 4%
- Japan: 3%
- Korea: 3%
- Other: 10%

*Source:* Author's calculations based on data from the European Patent Office, 2006.

## *The Role of International Coinvention in the ECA Region*

U.S. patent data confirm the importance of international coinvention as shown in the EPO data, both for the region as a whole and for individual countries. To place this in context, it is useful to compare recent invention trends for the ECA region to those of India and China. Figure 2.4 compares trends in U.S. patent grants for seven ECA countries, India, and China. The rationale for comparing ECA patenting activities to those of India and China is that the large ECA middle-income countries such as Russia, Ukraine, and to some extent Poland and Kazakhstan, are looking to India and China as benchmarks. Countries that have been through a transition—from central planning in China and a semi-socialist economy in India—are seen as interesting examples of overcoming the traditions of socialism. We show that just a few decades ago, ECA had more patents than India and China, but this has been changing in recent years.

The seven ECA countries compared are Russia, Hungary, the Czech Republic, Poland, Slovenia, Bulgaria, and Ukraine. The numbers are taken from the Cassis CD-ROM published by the U.S. Patent and Trademark Office at the end of 2006; the database incorporates data on U.S. patents granted up to the end of that year. Patent grants are attributed to the ECA region, India, or China, if any of the first 10 inventors listed on the patent reside in one of those countries, and these patent counts include patents granted to individuals, government agencies, and academic institutes, as well as firms.

**Figure 2.4. U.S. Patent Grants for the ECA 7, India, and China**

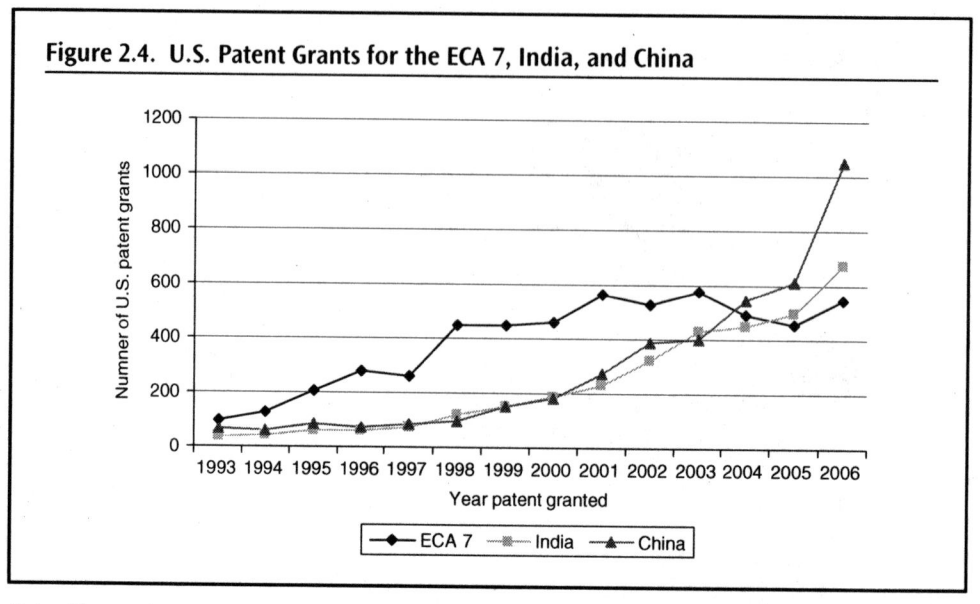

*Note:* The graph compares counts of patents in which at least one inventor is based in one of seven ECA countries, India, or China. The ECA-7 are Russia, Hungary, Poland, Slovenia, the Czech Republic, Bulgaria, and Ukraine.
*Source:* Authors' calculations based on the U.S. Patent and Trademark Office Cassis CD-ROM, December 2006 version.

Despite the impact of the transition period in the 1990s, inventors based in the seven ECA countries consistently received more U.S. patent grants than did inventors in India and China until the most recent years. If one sums up cumulative patent grants from 1993 through the end of 2006, the ECA 7 countries obtained 5,489 patent grants, whereas India-based inventors obtained only 3,331, and China-based inventors obtained 4,063. Clearly, when we normalize by population, the performance of the seven ECA countries has been much better on a per-capita basis. However, there is a clear acceleration in India- and China-based patenting in the most recent years, which is not evident in ECA patenting. Aggregate ECA patenting was slightly lower in 2006 than it had been five years earlier, but levels of patenting have been relatively stable. This contrasts sharply with the almost exponential growth in China-based inventive activity.

Figure 2.5 disaggregates ECA patent grants into those generated by international teams of inventors and those generated solely through the efforts of inventors based within a particular ECA country.

It is immediately apparent that international coinvention is *extremely* important—in recent years, more than half of total patent grants were generated from teams of inventors based in more than one country. While we do see some inventive collaboration between ECA countries, the patterns of coinvention are dominated by collaboration with inventors in more advanced countries. Germany, the United States, the other major European economies, Japan, and South Korea all tend to play more important roles than do other ECA countries. We also see significant and growing international collaboration in the U.S. patent grants of India and China, but it is much less prevalent than in the ECA countries.

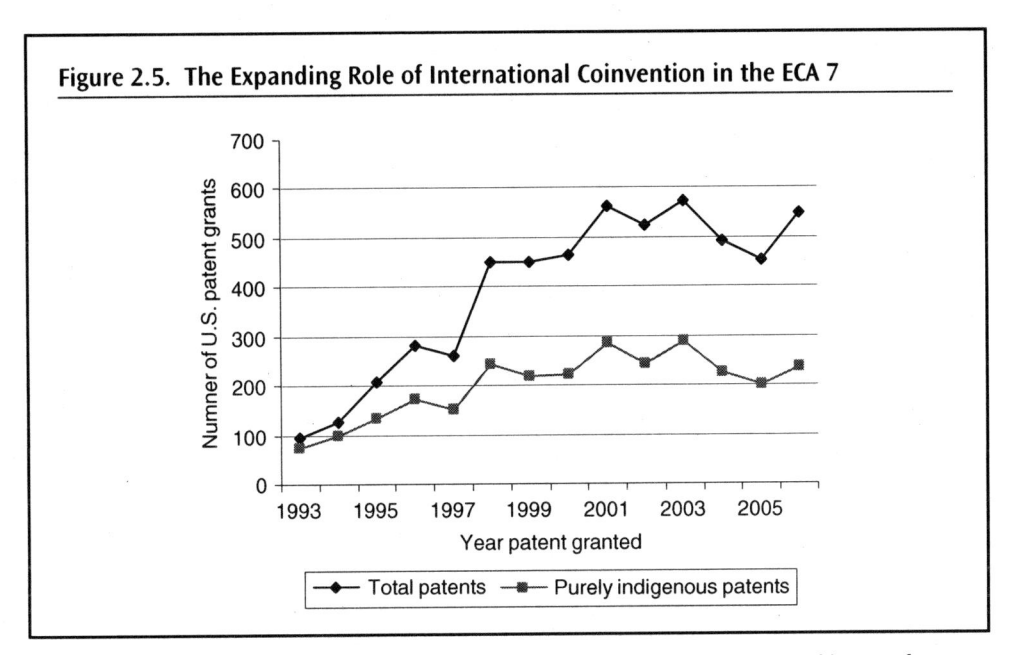

Figure 2.5. The Expanding Role of International Coinvention in the ECA 7

[Chart: Number of U.S. patent grants by year (1993-2005), showing Total patents rising from ~100 to ~550, and Purely indigenous patents rising from ~80 to ~250]

*Note:* The graph tracks total counts of patents in which at least one inventor is based in one of seven ECA countries: Russia, Hungary, Poland, Slovenia, the Czech Republic, Bulgaria, and Ukraine. "Purely indigenous patents" are those generated by a team whose members are all based in a single ECA country.
*Source:* Authors' calculations based on the U.S. Patent and Trademark Office Cassis CD-ROM, December 2006 version.

Our data reveal a much greater degree of international coinvention than do earlier analyses. In part, this is because much earlier work has been based on the National Bureau of Economic Research (NBER) Patent Citation Database, which ascribes the nationality of the patent to the location of the first inventor listed on the patent document. While incomplete, this method of assigning nationality generally "works" for patents generated in large, R&D-intensive economies like the United States or Japan. The overwhelming majority of U.S. and Japanese patents are produced by teams of inventors based entirely in the United States and Japan, respectively. However, the usual method of assignment is much less appropriate for smaller, less R&D-intensive economies; the method misses the striking increase in the importance of international teams of inventors, including those based in these smaller economies.

The work of other researchers using very different data also point to the strong role of multinationals in ECA-area invention. Violina Ratcheva, of the University of Sheffield in the United Kingdom, has utilized a data source called Locomonitor to track media announcements of R&D-oriented FDI in multiple regions, including the ECA.[7] In published papers and unpublished preliminary research, she finds evidence of strong concentrations of foreign R&D activity in the region. It is clear that the evidence of extensive

---

7. Locomonitor is a database of FDI projects maintained by OCO Consulting, a firm spun off from international accounting and consulting giant Price Waterhouse Coopers in 2001. Data are based on media and trade press reports.

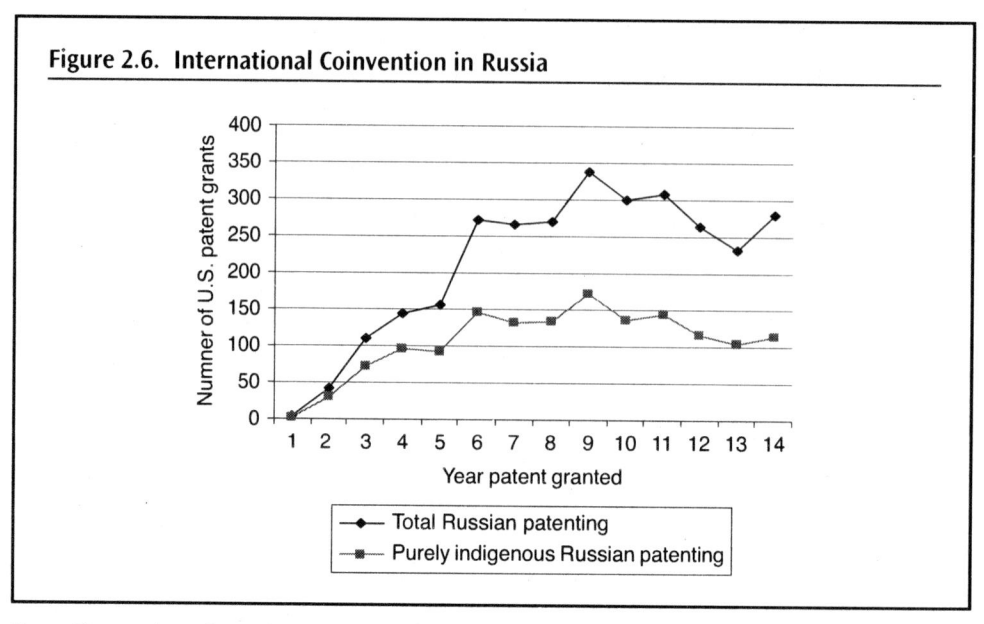

**Figure 2.6. International Coinvention in Russia**

*Note:* The graph tracks total counts of patents in which at least one inventor is based in Russia. "Purely indigenous patents" are those generated by a team whose members are all based in a single ECA country.
*Source:* Authors' calculations based on the U.S. Patent and Trademark Office Cassis CD-ROM, December 2006 version.

international coinvention that we document in this study is not some artifact of patent data but, instead, points to an economically significant real phenomenon.[8]

The high prevalence of international R&D collaboration evident in ECA patenting is likely to be a positive influence on the extent and nature of national inventive activity. Through collaboration with inventors based in more advanced industrial economies, ECA inventors are likely to achieve a greater understanding of recent technological trends and developments than would be possible through autarkic R&D effort. In the next section, we will view preliminary evidence based on U.S. patent citations consistent with this hypothesis. In fact, we believe our data support the tentative conclusion that the quantity and quality of ECA invention would be significantly lower in the absence of this collaboration.

Before turning to the patent citation analysis, we take one more look at international R&D collaboration within the biggest ECA economy: Russia. Figure 2.6 shows that the ECA regional trends hold for its single largest member state. A majority of Russia's U.S. patent grants are generated by cross-national inventor teams, and many of these patents are assigned to U.S. and other foreign multinationals. In fact, U.S.-based firms and other organizations generate about as many patents in Russia as do inventor teams composed solely of Russians.

---

8. For examples of recent research employing the Locomonitor database, see Demirbag, Ratcheva, and Huggins (2006) and Huggins, Demirbag, and Ratcheva (2007).

Table 2.1. Top 10 Russian Generators of U.S. Patents

| Rank | Name | Nationality | Number of U.S. Patents |
|---|---|---|---|
| 1 | LSI Corp. | United States | 111 |
| 2 | Samsung Electronics | South Korea | 71 |
| 3 | General Electric Co. | United States | 37 |
| 4 | Elbrus International | Russia | 36 |
| 4 | Sun Microsystems | United States | 36 |
| 5 | Ceram Optec Industries | Germany | 28 |
| 6 | Nippon Mektron | Japan | 26 |
| 7 | Ajinomoto Co. | Japan | 25 |
| 8 | Otkrytoe Aktsionernoe Obschestvo | Russia | 20 |
| 9 | Procter & Gamble | United States | 19 |
| 9 | Ramtech | United States | 19 |
| 10 | Advanced Renal Technologies | United States | 16 |
| 10 | Corning Inc. | United States | 16 |
| 10 | Nortel Networks | Canada/United States | 16 |

*Note:* The graph tracks total counts of patents in which at least one inventor is based in Russia.
*Source:* Authors' calculations based on the U.S. Patent and Trademark Office Cassis CD-ROM, December 2006 version.

Table 2.1 lists the top 10 creators of Russia-based U.S. patents. U.S. and Japanese multinationals figure prominently in this list. This top 10 list is somewhat misleading, however. There are a large number of German firms that collectively account for a large number of patents, even though each individual firm has registered relatively few.

## *Patterns of ECA Knowledge Absorption as Revealed by ECA Patent Citations*

We now turn to the use of patent citations as a window into the knowledge absorption process. Examination of the citations in indigenous patents from ECA countries taken out with the USPTO reveals some striking contrasts with those from other technologically successful regions. Indigenous patents typically make fewer citations to the preceding literature; they tend to cite older technologies, less fundamental prior inventions, and patents that are less frequently cited. All of this is suggestive of the notion that ECA inventors are insufficiently grounded in the recent technical state of the art. They are not sufficiently well-versed in recent technical developments outside of their region to build on those developments with maximum effectiveness. Case- and interview-based studies of R&D and productivity in the region have long criticized the tendency for centers of ECA scientific activity to be insufficiently connected to centers of technical excellence outside the region. Our data analyses, drawing upon data from thousands of individual patent grants from across the region, document citation patterns consistent with these criticisms.

This section of our research requires extensive data on patent citations, and the USPTO Cassis database does not contain citation mapping that links patents to the previously granted patents they cite. Therefore, we had to turn to other data sources. Professor Bronwyn Hall of the University of California, Berkeley, has undertaken a partial update of the well-known NBER patent citation database that is current for patents granted through 2002. In keeping with past conventions, nationality is assigned to patents on the basis of the address of the first inventor. We already know that this practice is imperfect for analyzing ECA patenting, but our use of the Hall database constrains us to use it in the analysis that follows. However, if it is the case that the first inventor is also the most important contributor, this convention would make this less of a limitation, as by this definition it would then capture patents in which ECA citizens have a greater contribution.

We are also constrained in the time dimension: The Hall update is only complete for patents granted through 2002. We therefore began our analyses by first identifying all U.S. patent grants over the 1990–2002 period for which the first inventor listed an address in one of the ECA countries. This produced slightly more than 3,000 granted patents. In order to compare ECA inventive activity with activity in other regions, we also sought to obtain a roughly equivalent number of patents from the following countries/regions: Emerging Asia (India and China), Advanced Asia (South Korea and Taiwan), Latin America (Brazil and Mexico), the European Periphery (Spain, Portugal, Finland, and Israel), and the European Core (Germany). As it turns out, the Latin American and Emerging Asian countries included in our comparator set did not generate quite as many patents as the ECA countries did. The other countries or regions generated far more. To maximize the comparability of these samples, we sought to match each ECA patent with another patent drawn from each of these five other regions that had the same primary patent class and grant year as the ECA patent. This was done to ensure that our comparisons primarily reflected differences in inventor nationality rather than citation differences across different technological fields. So we began our analyses with approximately 3,000 patents from four of the six regions, and fewer drawn from Emerging Asia and Latin America.

The next stage was to obtain detailed information on the patents cited by these sample patents. Some of our initial sample patents lacked usable citation data and were therefore dropped from subsequent analyses. The vast majority of patents could be connected to Bronwyn Hall's patent citation mapping, allowing us to examine and compare the backward citations patterns therein.

The next stage was to draw a meaningful distinction between "purely indigenous" inventive activity and multinational inventive activity. Recall our constraint—the Hall data update does not contain complete information on the locations of all inventors. We will thus be unable to distinguish adequately between teams of inventors that are all based in an ECA country and teams of inventors that include members based in advanced economies. However, as we have already noted, a relatively large fraction of patents with ECA inventors are assigned to U.S. firms or to large non-U.S. multinationals based outside the region. Some of these patents are created by the ECA-based subsidiaries of U.S. firms. Others are generated by nominally independent ECA firms undertaking research for U.S. clients under terms that assign the intellectual property rights to the U.S. client. When we include these multinational patents in our ECA sample, ECA patents do not appear to be that different from patents generated in other regions. However, when we omit these multinational patents from consideration, focusing on the "indigenous" patents and comparing them to indigenous patents in other regions, we find some striking differences between ECA-generated patents and those generated elsewhere.

In many cases, indigenous ECA patents make *fewer patent citations*, and the differences are often quite significant at traditional levels of statistical significance While the presence of a large number of citations is sometimes viewed as evidence that the citing patent is less innovative, we believe that we are correct in seeing the low levels of ECA citations as indicative of an insufficient grasp of global best practices and recent advances in the state of the art. This interpretation is strengthened by the existence of another empirical regularity: indigenous ECA patents typically cite older patents than do patents generated in other regions. It is hard to explain this on the basis of the age of the patent or technology class, but we started with a sample essentially matched in both aspects. We also see that indigenous ECA patents tend to cite patents with a less broad technological impact than those in other countries. Hall, Jaffe, and Trajtenberg have developed a measure of the generality of a patent's impact by looking at the distribution of citations received by that patent across different technology classes. Most patents are cited largely by other patents in the same or closely related patent classes. Only "fundamental" inventions tend to receive citations from a wide variety of patent classes. Patents cited by indigenous ECA patents tend to be systematically less "general" than other patents. They also tend to receive fewer citations over time. Because one would naturally think that older patents receive more citations, this difference is all the more striking. ECA citation patents only look "good" in comparison to those of the Latin American economies. That in itself is a rather striking statement about what these data show.

Statistical tests require us to measure the mean characteristics of patents cited by indigenous ECA patents with the mean characteristics of other groups. Once we focus solely on indigenous patents, the groups tend to be of different sizes, and we do not want to impose the assumption of a normal distribution for characteristics that tend to be highly skewed.[9] Brahmbhatt and Hu (2007) find similar results for patents in East Asia.[10] Trends in East Asia signal the growing intraregional dimension in East Asian knowledge flows.

---

9. We therefore conduct Mann-Whitney tests of equality (also known as the two-sample Wilcoxon rank-sum test), using STATA. We show the sample means (table A.1) and the p-value for the hypothesis that they are the same (table A.2). While small in absolute terms, the differences in means are not small in percentage terms, and imply test statistics well beyond the critical values in many cases. In 14 of 20 cases, the statistical test of equality is resoundingly rejected. At the 10 percent level, one can reject equality in 16 of 20 cases.

10. The United States is by far the largest source of citations for East Asian innovators, providing close to 60 percent of the total. Japan is the second-largest source, contributing close to 20 percent, on average. Further, they find that there is a rise of intraregional and compatriot knowledge flows (the share of citations by inventors in one East Asian economy to other patents in the same economy) for individual East Asian economies, indicating that the share of citations to other East Asian economies (typically to patents of Korea and Taiwan [China]) is highest-around 7 percent to 8 percent—in China, Hong Kong (China), Malaysia, and Singapore. On the other hand, the share of own or compatriot patents is highest in Korea (around 6 percent) and Taiwan (China), where it is over 10 percent. In other East Asian economies, Singapore shows an exceptionally high citation frequency to patents in Taiwan (China) and Korea, both of which significantly exceed (also high) citation frequencies to patents in Japan and the United States. Citation frequencies to Korea and Taiwan (China) in China and Malaysia also exceed those to Japan and the United States.

## Summary of Key Learnings

This section shows that much can be learned about potential pathways of knowledge absorption in the ECA region through the analysis of patent data. We see that the transition period of the early 1990s severely disrupted inventive activity and the kind of absorption of existing trends in the technological state of the art necessary for invention. As patent growth resumed, we see that international R&D collaboration became quite prevalent in the ECA region; this is evident in both EPO and USPTO patents. International coinvention has contributed significantly to the magnitude of ECA inventive activity.

Evidence drawn from USPTO patent citations suggests that international R&D collaboration has also contributed significantly to ECA invention quality. First, the long-standing criticism of "indigenous" ECA science and inventions as being insufficiently connected to the global technological mainstream appears to be largely correct. We see significant differences in ECA citation activity that are completely consistent with this view. Second, international R&D collaboration of various forms has allowed the ECA region to partially sidestep the handicap to its own R&D productivity imposed by its insufficient grounding in the global technological mainstream and by recent advances in the state of the art. When we factor back in the patents generated through this cooperation, the ECA citation patterns appear to be less distinctive. R&D cooperation helps to remedy shortcomings in ECA inventors' knowledge of global best practices.

## Implications for Policy

### The National Innovation Systems of CIS Countries Require Further Reform

The persisting effects of the legacy of Soviet R&D policies have been discussed widely, and the recent World Bank report *The Path to Prosperity: Productivity Growth in Eastern Europe and the Former Soviet Union* (2007a) outlines some of the remaining inefficiencies:

> Currently, many of the S&T (science and technology) resources are isolated both bureaucratically (in the sense that they are deployed in the rigid hierarchical system devised in the 1920s to mobilize resources for rapid state-planned industrial development and national defense), functionally (in the sense that there are few links between the supply of S&T output by research institutes and the demand for S&T by Russian or foreign enterprises), and geographically (in the sense that many assets are located in formerly closed cities or isolated science/atomic cities).

A disproportionate share of total R&D resources remains locked up in specialized state research institutes, in which there are no strong incentives to transfer commercially useful technology developed within the institutes to private entities in the ECA economies, which could put the technology to productive use.

Our results support these criticisms. Let us start by reviewing Table 2.1, listing the top 10 Russia-based organizations that have obtained patent grants in the United States, the world's largest and most important single patent jurisdiction. This list is dominated by the local research operations of foreign firms. Despite their absorption of a large share of Russia's total R&D expenditures, the state-led research institutes produce fewer patents than commercial companies. Surely, these research institutes and private commercial enterprises are not strictly comparable, as the type of research they do is very different. However, this does not detract from the main message (and even reinforces it), that resources going to these research institutes could be put to much more productive use by supporting R&D in the private sector. In Russia, the share of researchers in the population, and the aggregate outlays of R&D in GDP, are comparable to those of Germany and Korea, and are far ahead of those of Brazil, China, and India. But these high levels of inputs do not translate into high value added per capita, with Russia lagging behind the OECD, as well as other large middle-income countries in R&D outputs, along with a relatively low number of patents and scientific publications per capita (Schaffer and Kuznetsov in Desai and Goldberg 2008).

What is true in Russia appears to be true more generally throughout the CIS countries, but less so in Eastern Europe. The traditional institutions conducting research have been criticized for their isolation from international technological trends. We find statistical evidence of this relative isolation in the citation patterns of U.S. patents generated by indigenous inventors, when they are compared to comparable inventions of comparable vintage generated in other parts of the world. These indigenous patents generally make fewer citations to the existing state of the art, and they cite narrower inventions. The traditional institutions have been characterized by the poor quality of their inventions. We see evidence of the limitations of indigenous inventions in the relatively small number of citations that these patents receive from patents granted subsequently. Moreover, the number of indigenous patents is low relative to the level of R&D investment, and the number of patents is not sharply increasing, as they have been in countries such as India and China. All of these findings reaffirm the need for continued efforts to reform ECA R&D systems, and to complete the transition from the socialist-era model to a system modeled on global best practices that is more internationally integrated and market driven.

## *Foreign Firms are Making Significant Contributions to ECA Inventive Activity*

The research described above has documented the significant contribution that foreign firms have made and are making to ECA-region inventive activity. Foreign firms' local R&D operations, and their sponsorship of local inventors, collectively generate a large fraction of the total patents emerging from ECA countries. Not only does this process of international coinvention contribute to the quantity of ECA patents, but also raises the quality of ECA inventive effort. Whereas indigenous ECA patents lag behind other regions in terms of the degree to which they build on prior invention and extend it, the ECA patents created through multinational sponsorship are more effectively connected to global R&D trends and generally represent inventions of higher quality. Through collaboration with foreign scientists and engineers based in the world's innovation centers, ECA inventors are able to ground their efforts more solidly in the current technological state of the art, and to benefit from knowledge of recent, relevant technical developments outside the region. The multinationals—unlike the state institutes—have both the incentives for and structure to promote the translation of the ECA research effort into formal intellectual property that can then be deployed both inside and outside the region.

Interestingly, the large role of international coinvention is a feature of regional inventive activity that the ECA countries share with India and, especially, with China. As with the production of goods, it appears that ECA, India, and China all have the opportunity to benefit from participation in an emerging international division of inventive labor, in which local inventors become part of a "production chain" of knowledge.

The rapid growth of patenting in India and China presents a contrast to the lack of growth in the ECA region. A complete discussion of trends in R&D in India and China is well beyond the scope of this study, but we would like to point the reader's attention to one important similarity existing between these two countries and the ECA, as well as to one important difference.[11] Like the ECA countries, India and China have seen rapid growth in the share of their total patenting developed as a result of local R&D operations of multinational companies, or by international coinvention more generally. This is particularly true for China, where nearly half of U.S. patents granted to Chinese inventors were created by international teams of inventors, and only a part of these teams was actually based in China at the time of the patent application. The share of international coinvention in China's total U.S. patents is rising more rapidly than total patents, implying that international R&D collaboration has disproportionately contributed to the impressive growth in China's inventive output. The fraction of Indian patents developed as a result of international coinvention is lower, but is also experiencing rapid growth. As in the ECA countries, inventors in India and China are apparently able to overcome some of the weaknesses of their own evolving national innovation systems (NIS),[12] by participating in an international division of inventive labor that links them directly with centers of innovative excellence in Europe, Japan, and, especially, North America. Yet there is a very significant difference between ECA, on the one hand, and India and China on the other, and this is pointed out clearly in Figure 2.4 above. In recent years, ECA patenting in the United States has not grown significantly, while India and China have surged ahead. These differences reflect the following trends: extremely rapid economic growth in India and China; substantial public and private investment in higher education, and a large increase in the number of university-educated engineers and scientists; the much larger populations of the two giant Asian nations; and an increasing desire on the part of the world's leading firms to position themselves in these large consumer markets.

In a globalizing world, multinationals always have a choice among countries when it comes to locating R&D activity. ECA governments should encourage foreign R&D investment and international R&D collaboration. However, as shown in the literature review in ECAKE I, measures to support R&D are ineffective when these are poorly coordinated with other measures. An extensive literature has developed on the concept of a holistic national innovation system (the concept is elaborated in Nelson 1993; OECD 1998 and 2001; Lundvall and others 2002). The National Innovation System is a system in which those who generate new knowledge are efficiently connected to those who can benefit from its use. This connection is established through a set of instruments, institutional settings, and infrastructures that accelerate knowledge flow and enable innovation. For the system to work efficiently, the "links" form effective networks that help overcome market failures caused by coordination and information problems. This system, including the availability of

---

11. See Branstetter and Foley (2007) for a review of U.S. FDI in China, which includes a focus on the R&D activities of U.S. affiliates in that country.
12. See a discussion of the NIS in the last paragraph below and Figure 2.6.

**Figure 2.7. National Innovation Systems**

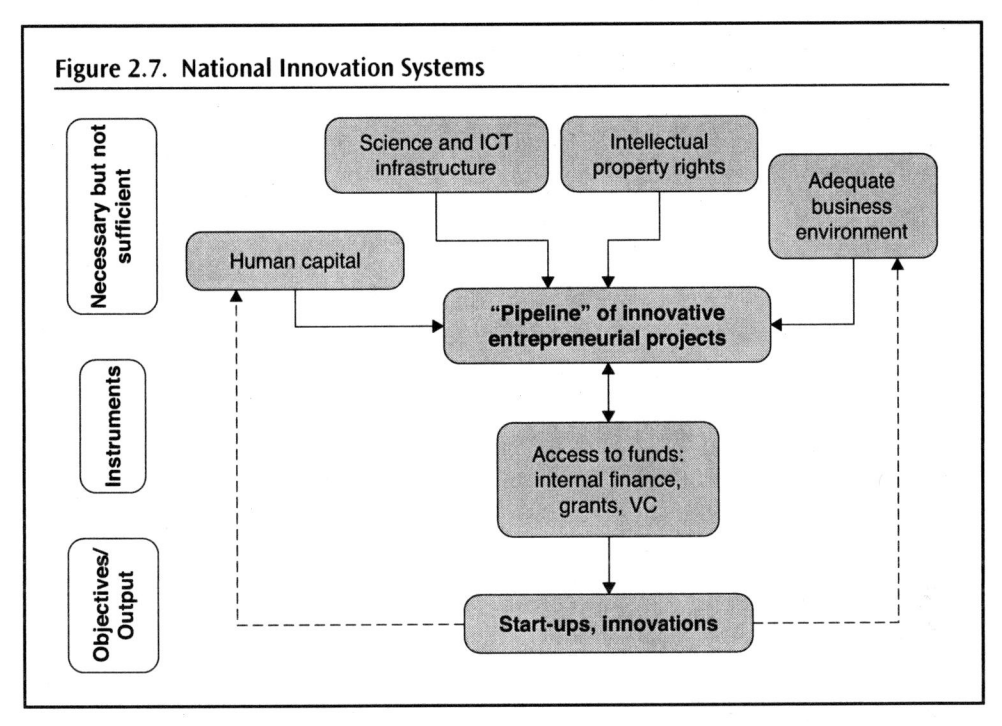

*Source:* Public Financial Support for Commercial Innovation, ECAKE I, Page 38.

human capital, science and ICT infrastructure, intellectual property rights (IPR), and a business-friendly environment, is illustrated schematically in Figure 2.7 which is reproduced from ECAKE I, page 38.

# CHAPTER 3

# The Links among Knowledge Absorption, Trade, and FDI

*By Lee Branstetter, Itzhak Goldberg, and Smita Kuriakose*

Openness to foreign trade and investment is critical to the process of technological absorption and diffusion, not only for the competitive pressure that it exerts on management and corporate governance, but also for the exposure to global best practices technology and management techniques that such openness provides to local firms. The contribution of international openness to growth has been evident since the first wave of globalization in 1870–1913. In his book, *The Mystery of Economic Growth* (2004), Helpman focuses on four themes:

1. Accumulation of physical and human capital,
2. Productivity growth, as determined particularly by the incentives for knowledge creation, R&D and learning by doing,
3. Knowledge flows across national borders and the impact of foreign trade and investment on incentives to innovate, imitate, and use new technologies,
4. Institutions that affect incentives to accumulate and innovate.

The second and the third themes are at the center of this study in general, and of this chapter in particular. In analyzing the effects of trade and FDI on technology absorption, we control for the effects of human capital and measures of investment climate.

This chapter focuses on the role of international trade (and participation in global supply networks) and FDI as mechanisms for the diffusion and absorption of technology in the ECA region. We consider trade and FDI together because they are two aspects of the singular phenomenon of "international connectedness." In fact, some of the measures of international connectedness that we employ in our analysis of the impact of international trade are at least as much a reflection of FDI as they are of the engagement in international trade.

We emphasize at the outset that we are interested in the impact of international trade, participation in global supply networks, and FDI as mechanisms to promote the diffusion and absorption of technology. We need to draw a distinction between this phenomenon and the narrower concept of productivity "spillovers," which has been the focus of much of the earlier literature. A "spillover" is a benefit for which the recipient did not have to pay. Previous researchers in the literature have often presumed that the technology absorption abetted by trade or FDI would show up in the form of faster growth rates of total factor productivity: Through contact with superior foreign technology, indigenous firms would be able to realize large gains in output that are far greater than their investments in new capital or better workers. Our broader concept of technology absorption or technology diffusion accepts the reality that indigenous firms may have to make costly investments to acquire new technical competencies, and their ability to appropriate the gains generated by these new competencies may be limited. In other words, a socially beneficial acquisition of new technological capabilities may take place, even when the conventionally defined "productivity gains" are limited or slow in coming, and it may take place even when the conventional techniques designed to measure "productivity spillovers" suggest limited effects. We will re-emphasize this distinction later in the chapter.

## Trade and its Benefits

Imports of manufactured goods, especially from more advanced economies, have been a channel of knowledge flows. By purchasing imported inputs and capital goods, firms in less developed countries acquire use of the technology embodied in these goods, thereby realizing gains in national welfare. Empirical multi-country studies (Coe and Helpman 1995; Keller 2002) show that international trade mediates flows of knowledge, allowing firms and industries to acquire technology that expands their productive capabilities, often in ways that show up in conventional productivity measures. There has long been a belief among experts that trading with countries that have a richer R&D stock or, more broadly, which are able to export more advanced technology goods, can facilitate the acquisition of new technical competencies on the part of importers. Conversely, one may posit that participation in export markets enables firms to become more productive, a phenomenon referred to as "learning through exporting," although this belief has not always been supported by empirical research at the firm level. In a similar vein, it is widely believed that trading in parts and components with foreign companies that are already well integrated in the global production network can facilitate the acquisition of new technology. Data from surveys of ECA firms provide direct evidence suggesting that the purchase of foreign capital goods is a major source of acquisition of newer, more effective technologies.[13]

Historical accounts of the rise of East Asian export industries, often based heavily on case studies and managerial interviews, stress the role played by advanced country buyers

---

13. In one question in the BEEPS dataset described below, ECA firms are asked to identify a single source (presumably the most important) of acquisition of new technology. The overwhelming majority cite the purchase of new capital equipment as the source of this new technology.

as conduits of technology and managerial know-how to developing country firms (Pack and Westphal 1986). Similarly, case studies and interviews suggest that the cultivation of manufacturing outsourcing relationships between local firms and multinationals—with or without a local presence in the market—can also serve as a channel of knowledge flows (Pack and Saggi 2001). These channels are likely to play a significant role in the ECA context. Our firm-level data, described below, allow us to measure, in principle, the impact of participation in such global supply networks on measures of the adoption of new-to-the-firm technology.

The significant, positive impact of openness to international trade on conventionally measured productivity growth, which has been found in the aggregate country-level studies mentioned above, has been confirmed at the firm level.[14] A study of trade liberalization and productivity in Chile (Pavcnik 2002), found significant positive effects of liberalization on within-firm productivity growth.[15] Other researchers have found similar effects in other contexts. An opening to trade typically intensifies the degree of market competition faced by local firms along a number of dimensions. Foreign firms are often both more efficient in production and better able to offer high-quality products than are many domestic incumbents. Their entry into the marketplace typically intensifies efforts on the part of domestic firms to improve productive efficiency and to alter the product mix so as to better compete with the new entrants. Efforts along both dimensions improve consumer welfare.

## Trade in ECA

As described in the study, *Productivity Growth, Job Creation, and Demographic Change in Eastern Europe and the Former Soviet Union*, merchandise exports and imports as a share of GDP in ECA expanded from just over 15 percent in 1994 to nearly 25 percent in 2003. Services, which were considered to be low priority under central planning, have emerged as a dynamic force in sectors such as telecommunications, transportation, energy services, and banking—and services trade as a share of GDP is 5 percent. These numbers, however, mask significant intra-regional variation: merchandise exports and imports, as a share of GDP, ranged in 2003 from 40 percent for the EU-8 through 15 percent for the low-income Commonwealth of Independent States (CIS). Services trade as a share of GDP ranged from around 4 percent for the CIS countries, where services are heavily burdened by regulation and protected from competition, to over 7 percent for the Southeast European (SEE) countries (World Bank 2007b).

Returning to the question of whether international trade mediates flows of knowledge, thus raising productivity, one might also ask: Does the generally positive

---

14. Yet, this is by no means a guaranteed result: Often-cited results suggesting the absence of a strong productivity effect in particular national contexts include the results of Harrison (1994) in Côte d'Ivoire and Haddad, De Melo, and Horton (1996) in Morocco.

15. De Loecker (2007) and Bernard, Redding, and Schott (2006) have shown that this general finding of a positive impact of trade liberalization on within-firm productivity growth remains even when one allows, theoretically and empirically, for the possibility of multiproduct firms and firm-specific demand shocks that affect firm-level markups.

view of the role of trade in enhancing the productivity of indigenous firms also hold in the ECA region? A number of recent studies strongly suggest that the answer is "yes." ECA's flagship trade study[16] "*of Eastern Europe and Former Soviet Union since the Transition*" argues that country participation in international trade has borne much fruit: significant improvements have been observed in the allocation of domestic resources, enhanced productivity, and more rapid and sustained growth. Earle and Estrin (2003) conclude that there is a strong, positive effect of increased import competition on the labor productivity of Russian firms. Djankov and Hoekman (2000) find that an opening to trade, *per se,* does have an impact on enterprise productivity, but that these effects tend to be concentrated among non exporters who, by virtue of their domestic orientation, have been largely cut off from international competition prior to trade reform. Taken together, these studies suggest that the trade liberalization undertaken by the ECA countries in the 1990s did indeed have a positive impact on productivity.

Yet many countries in ECA still have high average tariff and non-tariff barriers, which would need to be reduced in the medium term in order to gain from international integration. In Russia, in particular, the average tariff increased between 2001 and 2003 from about 11.5 percent to between 13 percent and 14.5 percent, but it held steady in 2004 and 2005 (Shepotylo and Tarr 2007). This places Russia's tariffs at a level slightly higher than those of other middle-income countries, and considerably higher than the OECD countries. The trade weighted standard deviation of the tariff approximately doubled from 9.5 percent in 2001 to 18 percent in 2003, but then fell to 15.2 percent by 2005. Further, the authors suggest that the Russian tariff structure has become more diverse, implying more highly protected sectors and more sectors with very low tariffs. While Russia implemented a tariff simplification reform in 2000–01, wherein it significantly reduced the number of ad valorem tariff rates employed, there was a movement to a less uniform tariff structure. A diverse tariff structure, as argued by the authors, typically imposes significant inefficiency costs on a country, due to increased incentives for lobbying and rent-seeking behavior, which typically lead to tariff-setting policies that are highly inefficient.

A reduction in the import tariff by 50 percent would produce gains for the Russian economy on two counts. It would improve domestic resource allocation due to a shift in production to sectors where production is valued higher, based on world market prices; and it would increase productivity as a result of Russian businesses' ability to import modern technologies (Rutherford and Tarr 2006)—which is the more significant impact. Having established that tariff and non-tariff barriers pose obstacles to trade openness, Kee, Nictia, and Olarreaga (2006) compute indicators of trade restrictiveness (Table 3.1) that include measures of tariff and non-tariff barriers for 91 developing and industrial countries. Among the indices that the authors calculate is one that focuses on the trade distortions imposed by each country on imports, and another that focuses on

---

16. Introduction to *"Trade Record of Eastern Europe and Former Soviet Union since the Transition,"* Part I, page 51.

market access for exports in the rest of the world. Table 3.1 shows these two indices for the manufacturing sector. It is interesting to note that the Trade Restrictiveness Indices (TRI) for imports for Russia and Kazakhstan are lower than those of Brazil and India; they are, however, higher than those of South Africa, China, and the EU. Among the ECA countries for which Kee et al. estimated a TRI, Russia's is the highest—a reflection of the high tariff and non-tariff barriers that the country imposes on its imports. This fact is illustrated in Figure 3.2, where Russia has the highest overall trade restrictive index (OTRI)[17] for tariff and non-tariff barriers. However, it experiences smaller trade distortions on its exports from the rest of the world, with China the only country facing a lower level of restrictiveness. Romania has the highest overall trade restrictiveness index for tariffs alone (Figure 3.1), a reflection of the fact that the average Most Favored Nation (MFN) tariff in Romania is the highest in the region, at 11.6 percent, and both agricultural and manufacturing tariff protections are among the highest in Romania (World Bank 2005).[18] Russia imposes the highest non-tariff barriers, making it the most overall trade restrictive country in ECA (Figure 3.2). Early on in the transition, many countries in the region cut tariffs and reduced or eliminated non-tariff barriers. Governments in the region have taken the initiative to ease policy restrictions at their borders to facilitate trade flows. The weighted average applied tariff rate on all goods for all of the region's countries is 5.8 percent (World Bank 2005). These tariff rates compare favorably with

**Table 3.1. Trade Restrictiveness Indices**

| Country | Imports | Exports |
|---|---|---|
| Albania | 0.07 | 0.139 |
| Hungary | 0.073 | 0.086 |
| Kazakhstan | 0.123 | 0.115 |
| Kyrgyz Republic | 0.033 | 0.116 |
| Lithuania | 0.048 | 0.161 |
| Latvia | 0.067 | 0.114 |
| Romania | 0.16 | 0.116 |
| Russia | 0.19 | 0.078 |
| Poland | 0.07 | 0.108 |
| Ukraine | 0.183 | 0.106 |
| Brazil | 0.217 | 0.081 |
| China | 0.118 | 0.06 |
| India | 0.203 | 0.131 |
| South Africa | 0.066 | 0.074 |
| EU | 0.081 | 0.085 |

*Source:* Kee, Nicita, and Olarreaga 2006.

---

17. OTRI (tariff), displayed for selected countries in Figure 3.1, only summarizes the tariff restrictiveness of a country. It is the uniform equivalent tariff that maintains the aggregate import volume of a country at its current level with heterogeneous tariffs. On the other hand, OTRI (tariff + non-tariff), seen in Figure 3.2, summarizes the trade policy stance of a country. It is the uniform equivalent tariff that maintains the aggregate import volume of a country at its current level with heterogeneous tariff and non-tariff measures.

18. Average most favored nation (MFN) tariffs for countries in transition range from a low simple average of 3.3 percent in Armenia to a high of 11.5 percent to 11.6 percent in Belarus and Romania. The highest tariff protection in agriculture is present in Bulgaria and Romania, while the highest tariff protection in manufacturing is in Belarus, Romania, and Uzbekistan (World Bank 2005).

**Figure 3.1. Overall Trade Restrictiveness Index-tariff (all goods), 2006**

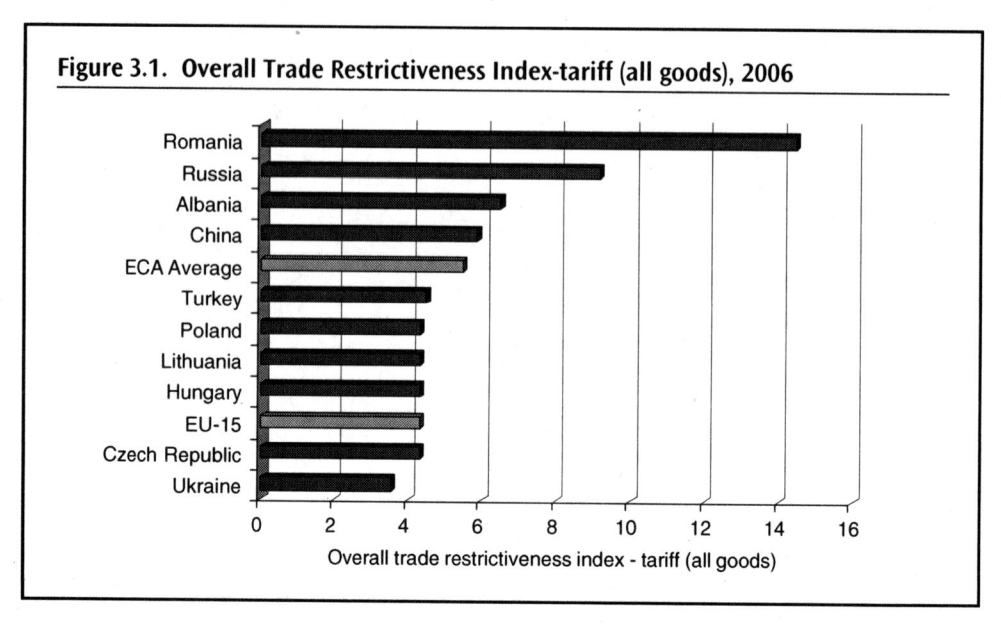

[Bar chart showing values from highest to lowest: Romania, Russia, Albania, China, ECA Average, Turkey, Poland, Lithuania, Hungary, EU-15, Czech Republic, Ukraine. X-axis: Overall trade restrictiveness index - tariff (all goods), range 0 to 16.]

*Note:* The methodology used is from Kee, Nicita, and Olarreaga (2006).
*Source:* World Bank trade web site.

those of developing countries at comparable income levels. Nonetheless, some countries in the region still maintain high tariffs and appreciable non-trade barriers. These include some Central Asian republics and others in the CIS, such as Kazakhstan and the Kyrgyz Republic.

**Figure 3.2. Overall Trade Restrictiveness Index-tariff + non-tariff (all goods), 2006**

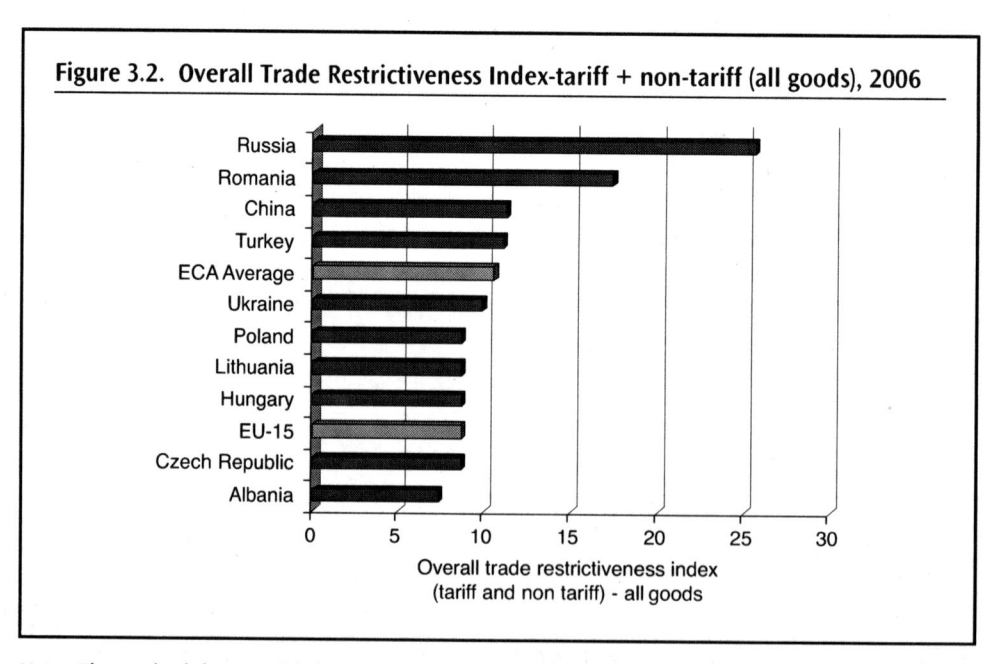

[Bar chart showing values from highest to lowest: Russia, Romania, China, Turkey, ECA Average, Ukraine, Poland, Lithuania, Hungary, EU-15, Czech Republic, Albania. X-axis: Overall trade restrictiveness index (tariff and non tariff) - all goods, range 0 to 30.]

*Note:* The methodology used is from Kee, Nicita, and Olarreaga (2006).
*Source:* World Bank trade web site.

Research has shown that improved and simplified customs procedures would have a significant positive impact on trade flows, and that the facilitation of cross-border movement of goods would have a positive effect on the ability of a country to attract foreign direct investment and better integrate into international production supply chains. These are the conclusions of the survey by Engman (2005). More specifically, Wilson, Mann, and Otsuki (2003) have shown that enhanced port efficiency has a large and positive effect on trade flows in the Asia-Pacific region and that regulatory barriers deter trade. Improvements in customs and greater use of e-business significantly expand trade, but to a smaller degree than do improvements in ports or regulations. Persson (2007) uses data from the World Bank's Doing Business Database on the time required to export or import, as indicators of cross-border transaction costs. Using a gravity model, she robustly finds that, on average, time delays both on the part of the exporter and the importer, significantly decrease trade flows.

Further, technical barriers to trade (TBTs) also significantly reduce the entry of firms into export markets, substantially reduce trade in some cases, and increase firm start-up and production costs. Chen et al (2006) have shown that testing procedures and lengthy inspection procedures significantly reduce exports and impede exporters' market entry.[19]

Because trade and FDI are the key channels for international diffusion of knowledge, poor logistics may impede access to new technology and know-how, and in turn slow the rate of productivity growth (according to the 2007 Logistics Performance Index).[20] Cost and efficiency of transport services determine export competitiveness, with higher shipping costs feeding into lower returns to capital in export ventures, and poor logistics increasing inventory costs (Mattoo, Rathindran, and Subramanian 2001). The Logistics Performance Index (LPI),[21] as shown in Table 3.2, analyzes the performance of key trade logistics processes for 150 countries.

It shows that ECA countries like Russia, Kazakhstan, Albania, and Tajikistan rank low at 99, 133, 139, and 146, respectively. The LPI shows that landlocked countries in Central Asia are typically the most logistically constrained. Further, there appears to be a negative correlation between the LPI and the Trade Restrictive Index (TRI), which would imply that policies to decrease trade logistics costs would also translate into a decrease in the degree of trade restrictiveness. Countries doing fairly well in logistics are also likely to do well in growth and competitiveness, export diversification, and trade expansion. The LPI (2007) shows that developing countries where trade has been an important factor in accelerating

---

19. They find that firms that outsource components are more challenged by compliance with multiple standards. Wilson and Otsuki (2001) have shown that adopting international guidelines regarding the pesticide aflatoxin would substantially increase world trade of nuts and cereals. Maskus, Otsuki, and Wilson (2005) have shown that standards and technical regulations raise the start-up and production costs of firms in developing countries to an extent that can be substantial for many companies.

20. Connecting to Compete, Trade Logistics in the Global Economy, The Logistics Performance Index and its Indicators, 2007.

21. For details, look at http://www.gfptt.org/lpi: The LPI is based on several areas of performance, such as trade procedures, infrastructure, services, and reliability. Based on a 1 to 5 scale (lowest to highest), it aggregates more than 5,000 country evaluations by professionals trading with the country on various dimensions of performance.

Table 3.2. Logistics Performance Indices, 2007

| Country | Logistics Performance Index Rank (Of 150) | Score (Max: 5.00) | Country | Logistics Performance Index Rank (Of 150) | Score (Max: 5.00) |
|---|---|---|---|---|---|
| Germany | 3 | 4.1 | Lithuania | 58 | 2.78 |
| United Kingdom | 9 | 3.99 | Brazil | 61 | 2.75 |
| United States | 14 | 3.84 | Croatia | 63 | 2.71 |
| Spain | 26 | 3.52 | Ukraine | 73 | 2.55 |
| Portugal | 28 | 3.38 | Belarus | 74 | 2.53 |
| China | 30 | 3.32 | Bosnia and Herzegovina | 88 | 2.46 |
| Turkey | 34 | 3.15 | Macedonia | 90 | 2.43 |
| Hungary | 35 | 3.15 | Russia | 99 | 2.37 |
| Slovenia | 37 | 3.14 | Kyrgyz Republic | 103 | 2.35 |
| Czech Republic | 38 | 3.13 | Azerbaijan | 111 | 2.29 |
| India | 39 | 3.07 | Serbia and Montenegro | 115 | 2.28 |
| Poland | 40 | 3.04 | Uzbekistan | 129 | 2.16 |
| Estonia | 47 | 2.95 | Armenia | 131 | 2.14 |
| Slovak Republic | 50 | 2.92 | Kazakhstan | 133 | 2.12 |
| Romania | 51 | 2.91 | Albania | 139 | 2.08 |
| Bulgaria | 55 | 3.92 | Tajikistan | 146 | 1.93 |

*Source:* Connecting to Compete, Trade Logistics in the Global Economy, 2007, http://www.gfptt.org/lpi, The World Bank.

growth[22] also have logistics that are significantly better than in other countries at similar income levels. It also goes on to suggest that in emerging economies, where export-oriented manufacturing has been a major factor in growth, the private sector has been a key proponent of logistics reforms.

## *Is There Learning by Exporting?*

In the reviewed literature, the concept of "learning by exporting" has been seen as a process in which exporting increases productivity by exposing producers to new technologies, or by upgrading product quality. Exporting is another channel through which firms based in open economies can acquire foreign knowledge about technologies and products.

---

22. The study uses multivariate regressions and shows that there exists significant association between LPI and outcomes such as medium-term growth from 1992–2005 and trade expansion, defined as the overall annual change in trade openness over the same period.

Both case studies and empirical evidence support this view, showing that exporting firms are more efficient than non exporting firms (Pack 1993).

Because import protection can also restrict the growth of exporting firms and industries, trade liberalization (reducing tariffs and removing non-tariff barriers) is often also associated with export growth. The importance of exports for absorption has been long debated in the literature. Some empirical studies have found evidence of "learning by exporting"—that domestic firms are able to improve their technology of production through exports to more advanced markets (Van Biesebroeck 2005; Kraay, Soloaga, and Tybout 2002). However, this has not proven to be a robust finding. The apparent absence of learning-by-exporting effects documented in many contexts has led many economists to question their importance and has, to some extent, undermined the policy case for pro-trade reform. Even well-cited papers such as Pavcnik (2002), which do find evidence of productivity gains from trade liberalization, tend to find most of these gains arising through increased import competition, and have found no evidence of learning by exporting. This would seem to suggest a limited relationship between trade liberalization and the increased absorption of advanced technology by exporting firms.

Our study strongly suggests that the conventional methodology employed by virtually all previous researchers to estimate learning-by-exporting effects may be biased against finding a linkage between exports and increased technology absorption, even when such a relationship exists. This bias exists, in part, because the conventional methodology uses changes in total factor productivity (TFP) as a proxy for technology absorption and diffusion. We will demonstrate below how this methodology might fail to find evidence of efficiency-enhancing technology absorption, even when it does exist. As the reader will see, a similar critique may apply to previous attempts to estimate the impact of foreign direct investment on technology absorption.

## Foreign Direct Investment

The close connection between foreign direct investment and trade must be stressed. ECA's flagship trade study *From Disintegration to Reintegration: Eastern Europe and Former Soviet Union in International Trade* (World Bank 2005) emphasizes this connection and places it in a theoretical and global context. The benefits of international exposure to best practice technologies often come directly or indirectly through the intermediation of foreign firms. The strength of this connection arises out of well-noted revolutionary changes in logistics, information technology, and manufacturing in recent decades, which allow firms to disaggregate the total production process of a good or service into separate stages, and to locate each of these stages in an environment where local factor endowments enable efficient production. This process, referred to as the slicing up of the value chain, or the fragmentation of production (Krugman 1995; Jones, Kierzkowski, and Lurong 2004), enables a broad range of countries to participate in various stages of the production chain—even if they do not possess intangible assets or a service infrastructure sufficient to provide them with a comparative advantage in the production of final goods. Multinationals have played an important role in driving this process—estimates suggest that about two-thirds of world trade by the latter half of the 1990s involved multinational corporations.

Multinationals have played particularly significant roles in expanding the international trade in some of the most successful developing countries. China clearly stands out over the course of the past decade, in terms of the speed with which its exports of manufactured goods have expanded and diversified across a broad range of product categories. China is now the third-largest trading economy, as measured by conventional trade statistics, and it has recently emerged as the largest single supplier of IT hardware to the U.S. market. Most of the expansion of Chinese trade in the 1990s has been driven by foreign firms, and this is especially true of "high-tech" production. Today, foreign firms are responsible for nearly 60 percent of China's total exports. By 2003, foreign firms accounted for 92 percent of China's export of computers, computer components, and peripheral devices.[23]

China is not alone in Asia in terms of its reliance on foreign firms to mediate exports of technologically intensive goods. In fact, China can be viewed as merely the latest manifestation of a phenomenon that is common throughout Southeast Asia. Athukorala (2006) presents an interesting view of Asian industrial history, in which he contrasts the industrialization of Japan and the East Asian Newly Industrialized Economies (NIEs), which were primarily based on the growth of indigenous firms, with the experience of "latecomer countries"—the Southeast Asian nations and China. He argues that a long period of import substitution in the latecomer countries undermined the development of indigenous entrepreneurship and cut off local managerial elites from international markets. In this context, it was simply more efficient for the latecomer countries to join a preexisting international production chain, rather than try to create one from scratch through the efforts of indigenous firms alone. Using official data from Asian countries to document the sizable expansion in the share of international trade mediated by multinationals in all of these countries, he concludes that the entry of multinationals is virtually essential for the export success of latecomers.

In ECA, seven countries—the Czech Republic, Estonia, Hungary, Poland, the Slovak Republic, Slovenia, as well as Turkey—started off by participating in network trade in clothing-a sector activity intensive in unskilled labor and embedded in so-called "buyer-driven" production chains, wherein global buyers create a supply base upon which production and distribution systems are built, without direct ownership. However, prompted by rising wages, these countries were able to move into sectors such as automotive and information technology (IT)—which are intensive in skilled labor and capital and, in the case of IT, intensive in knowledge as well—by using foreign capital and know-how and the so-called "producer-driven" production chains. The latter networks divide the value chain into smaller components and move them to countries where production costs are lower. Production fragmentation in vertically integrated sectors (drives) producer-driven network trade. Such a transition to producer-driven networks has, however, not taken place for CIS countries such as the Kyrgyz Republic, Moldova, and Turkmenistan, which are involved in the buyer-driven production chains of the clothing trade. Nor has it happened in Belarus, Russia, and Ukraine which, together with the Kyrgyz Republic and Moldova, have been active in the furniture network; though buyer driven, the furniture network is

---

23. See Branstetter and Lardy (2007) for an overview of Chinese trade and FDI reform, and which stresses the role of foreign firms in China's export dynamism.

more diversified and complex, requiring a larger input of skills and investment in capital assets and thereby creating more opportunities for knowledge transfer and productivity spillovers compared to the clothing network.

Proximity to Western Europe has evidently been a major advantage with regard to countries' participation in networks. Indeed, the EU-15 accounts for three-quarters of the producer-driven network trade of the transition countries and Turkey as a whole. Notwithstanding the disadvantages of geography, reform of the investment climate, which is within the control of country authorities, is important in order to attract the FDI necessary to permit participation in such trade. The three most important obstacles to doing business are reported to be: (i) the macroeconomic environment, (ii) regulation, especially customs and trade regulations, and (iii) the judiciary and corruption in institutions. Hence, reform in these areas is a high priority for policy reform in the low-income countries of the region. We will come back to this point in the last chapter on the role of government (World Bank (2007b).

## *FDI and Technology Absorption*

We now return to the question of how and to what extent FDI has contributed to technological absorption—both in general and in the ECA region in particular. Empirical studies[24] generally find strong, robust effects of foreign ownership on the productivity of wholly or partially owned foreign subsidiaries. In other words, firms with some level of foreign ownership are more productive and technology intensive than are purely domestic firms. What is rarer is evidence of productivity spillovers to indigenous domestic firms. A number of studies have found evidence of such spillovers in industrialized countries, but they have proven harder to find in developing-country contexts. In fact, a number of well-regarded studies have even found evidence of negative TFP spillovers, suggesting that the presence of foreign firms actually triggers a *decline* in the productive efficiency of local firms. Studies of the Czech Republic, Bulgaria, Romania, and Poland could not find evidence of positive spillovers to indigenous firms (Djankov and Hoekman 2000; Konings 2000). Javorcik (2004) has suggested that spillovers from FDI will only flow to a particular subset of indigenous firms. She notes that multinationals have little incentive to transfer technology to local rivals, and can often takes steps to limit the extent to which their technology leaks out. However, they do have an incentive to share technology with local suppliers—and Javorcik finds evidence of inter-industry (or "vertical") spillovers from foreign firms through supply chain relationships to their local suppliers in Lithuania. Garrick Blalock (2002) finds similar evidence of spillovers through so-called backward linkages in Indonesia.[25]

---

24. This literature has taken a production function-based approach to estimate the impact of FDI on the productivity of both foreign subsidiaries and indigenous firms. Output levels or growth rates of output at the firm level are regressed on measures of inputs (or neoclassical assumptions are imposed on data to subtract cost share weighted measures of input growth from output growth), allowing the computation of a residual, which is interpreted as total factor productivity. The presence of FDI spillovers are inferred from the positive correlation, if any, between measures of multinational presence in the industry or related industry, and the productivity residuals of local firms.

25. Yet, recent research for China fails to find strong evidence for spillovers through vertical linkages (Girma and Gong 2007).

> **Box 3.1: Measuring the Impact of Trade and FDI on Technology Absorption**
>
> The conventional methodology estimating the effect of FDI on productivity cannot fully capture the impact of FDI on technology absorption by local firms, especially among firms within the same industry. Nor does it capture the full benefits from learning by exporting. This is because the methodology does not distinguish effectively between changes in profits and changes in the technical efficiency of production. When foreign firms arrive, or when firms are first exposed to international competition through trade liberalization, it is likely that this change will diminish the market power of the indigenous incumbents, driving down profits—at least in the short run. At the same time, the incumbents may learn useful production techniques and technologies from their foreign rivals or foreign customers, allowing them to reduce their costs and produce goods and services much more efficiently. The decline in profits arising from this learning could overshadow the increase in productivity in the data, even when the latter effect is quite large. The analysis below provides a graphical illustration of this problem.
>
> Most econometric studies cannot track changes in the prices charged by individual firms. Instead, the output of all firms in an industry is deflated by a common price index. As a consequence, the decline in profits following the entry of foreign firms through exports or direct investment, evident in Figures 3.3 and 3.4 (see Box 3.2), shows up as a decline in total factor productivity under the conventional methodology. This can be true even in cases, such as the one we have illustrated, where real learning by domestic firms takes place, resulting in quite substantial reductions in marginal cost. It may be that a decade of applied research has systematically underestimated the effects of FDI and openness on the technology absorption of local indigenous firms.
>
> The potentially serious shortcomings of the conventional approach point to the need for an alternative, and we provide one in the following section. By employing survey data, we can more directly measure the technology absorption of ECA firms. We find a strong empirical connection between technology absorption and the exposure of ECA firms to foreign markets and foreign competitors through trade and FDI. These results considerably strengthen the case for trade and FDI openness.

## *Empirical Analysis: Linking Trade, FDI, and Human Capital with Technology Absorption*

Having extracted significant lessons for the ECA region from the existing literature, we go on to use original data analysis to quantitatively assess their relevance in the ECA context. The BEEPS datasets contain a number of variables describing outcomes closely related to technology absorption, each of which is obtained from detailed surveys of managers of ECA-region firms. In these surveys, firm managers are asked specifically whether their firm recently introduced a new (to the firm) product, upgraded an existing product, acquired a new production technology, signed a new product licensing agreement, or acquired a new quality certification. Each of these potentially represents a dimension of the kind of technology absorption process that we believe is fostered through exposure to international best practices. In our empirical work, we use these variables, individually and together, to attempt a systematic assessment of the degree to which ECA-area firms really are absorbing technology through their connections to the global economy specified in the academic literature.

We are able to measure this connection because the BEEPS data also contain information on the extent of international "connectedness" of the individual firms. These data include information on exports, the level of foreign ownership, whether or not the firm is

## Box 3.2: New Foreign Competition: Lowering Profits and Raising Efficiency

This point is important enough that it deserves some elaboration, and a graphical illustration may be useful to underscore the idea. Imagine a local firm with a substantial degree of *de facto* monopoly power, perhaps held as a consequence of an only partly reformed trade and FDI regime. This firm's *ex ante* position is illustrated in Figure 3.3. The monopoly power of the local firm is represented by the relatively inelastic downward-sloping demand curve it faces. Responding optimally to this market power, the firm keeps output levels low and output prices high, charging the optimal monopoly price.

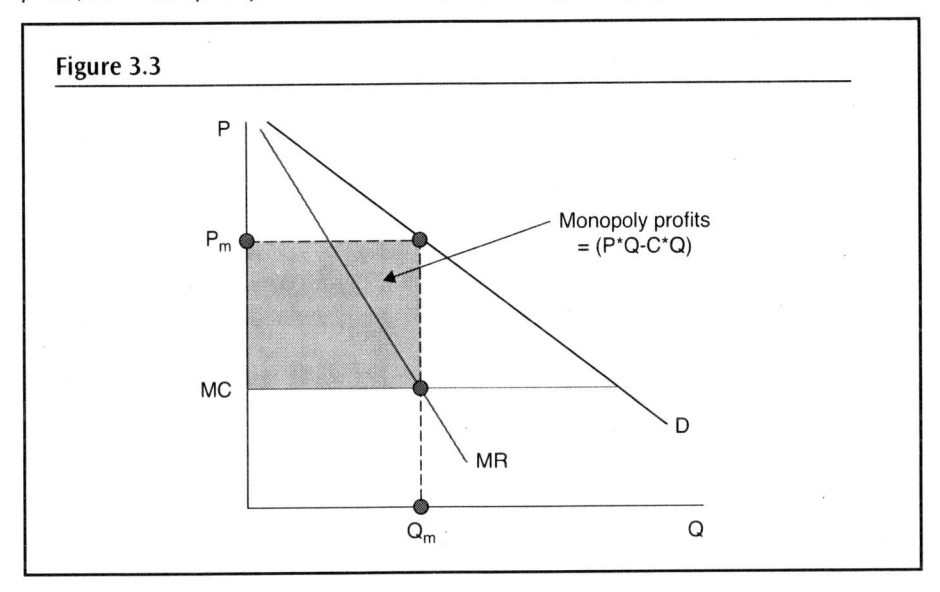

**Figure 3.3**

Monopoly profits = $(P^*Q - C^*Q)$

Now, thanks to trade and FDI reform, a foreign entrant comes into the market—an entrant with superior technology, higher quality, a more powerful brand name, or perhaps all three attributes. The local firm finds that its demand curve has both shifted in and become more elastic—that is, flatter. This is illustrated in Figure 3.4. The opportunity cost of foregone income at all price points is now higher. The local firm reacts rationally, lowering price and output below the *ex ante* monopoly levels.

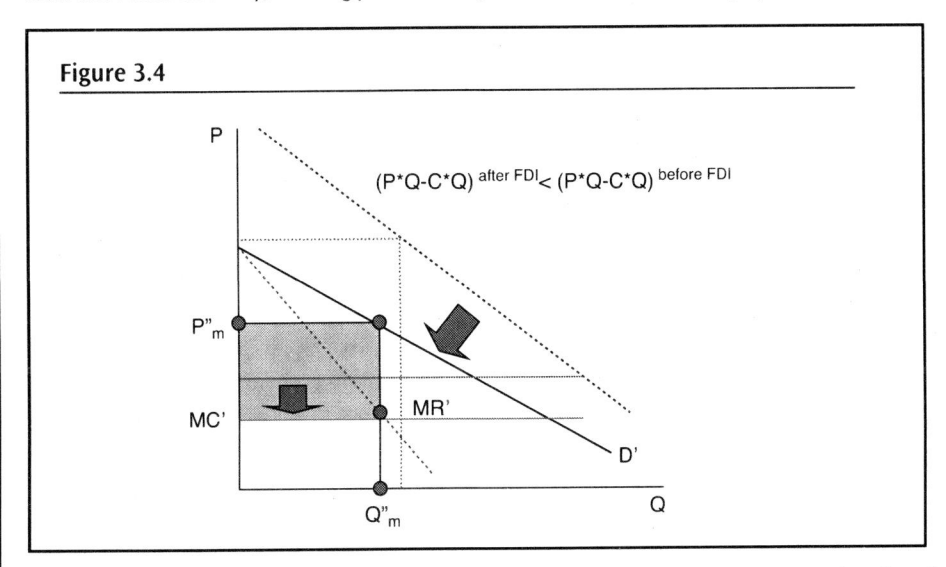

**Figure 3.4**

$(P^*Q - C^*Q)^{\text{after FDI}} < (P^*Q - C^*Q)^{\text{before FDI}}$

*(continued)*

> **Box 3.2: New Foreign Competition: Lowering Profits and Raising Efficiency** *(continued)*
>
> With the passage of time, the local firm is able to learn elements of the foreign competitor's technology, thereby realizing substantial declines in marginal cost. This allows the local firm to expand output and increase profits, relative to its situation immediately after the foreign firm's entry. Nevertheless, the profits of this firm remain lower than they were under the prior *de facto* monopoly. As emphasized in the text, the conventional methodology will often interpret these lower profits as indicating a decline in technical efficiency, even though the cost of production is now much lower.

a supplier to a multinational, how much it sells to multinationals, and whether the firm engages in joint venture partnerships with multinationals. It is thus possible, at least in principle, to measure the impact of international connectedness along various dimensions on the likelihood of technology absorption, measured in different ways.

To be concrete, we will consider regressions of the following form:

$$Upgrade_{it} = \beta_0 + \beta_1 \left(\frac{K_{it}}{L_{it}}\right) + \beta_2 International_{it} + \beta_3 Human\_Capital$$

$$+ \beta_4 Investment\_Environment_{it} + \theta_j + \gamma_c + \lambda_t + \varepsilon_{it} \qquad (1)$$

Where various measures of technology upgrading or technology acquisition are regressed on firm-specific (and sometimes time-varying) measures of capital intensity, the various measures of international connectedness described above, industry dummies (indexed by j), country dummies (indexed by c), and time dummies (indexed by t). Based on prior World Bank research, we know that the level of human capital within the firm and the institutional features of the home market that reward or penalize investment will also impact technology upgrading and technology acquisition. We will include these variables in some specifications to ensure that our key findings are robust.

The BEEPS datasets are cross-sectional surveys, and the most complete of these surveys is the most recent one, conducted in 2005. For this reason, much of our empirical work will focus on the data contained in this survey. However, an obvious identification problem arises in the context of cross-sectional data. Measures of international connectedness and measures of technology absorption both reflect choices by the firm, and both could be driven by some unmeasured or poorly measured third factor. In particular, we worry that firms with better managers will tend to be more proactive in terms of their adoption of technology; a cross-sectional regression is unable to determine whether positive correlations observed between these variables are driven by a causal relationship. If we had a good instrument that was highly correlated with firm-specific measures of international connectedness, but was not itself subject to similar endogeneity critiques, we could employ such an instrument. However, as is often the case, such an instrument does not appear to exist. In the absence of a good instrument, we will utilize panel data techniques to address these issues of causality, and we will further specify our approach below.[26]

Our contribution lies in establishing a direct connection between the exposure to international technical best practices, through trade and supply networks, and specific, discrete processes of technical improvement at the firm level. The prior literature has been unable to make much progress on this front, in part because of the difficulty of tracking technological absorption through the usual sorts of data available to researchers. The unique resources of the BEEPS surveys provide an unusual, if limited, ability to peer inside the "black box" of technology absorption, and those resources are exploited in this chapter.

The qualitative and subjective nature of the data on technology absorption present challenges to our empirical analysis. The notion of exactly what constitutes a "new" technology or a "new" product will clearly vary across countries, industries, and even individual firms within an industry. It seems evident from related work using these data that the typical respondents frame their answers about "new" products, processes, and technology with respect to the experience of the surveyed firm, rather than global best practices.[27] Thus, these variables are not capturing innovation, *per se*, as much as technology absorption; as such they are actually well-suited to our purposes. Because the frames of reference of multinational subsidiaries with regard to new technology may be quite different from that of indigenous firms, we need to be careful in our interpretation of results that compare majority or wholly owned foreign subsidiaries with other kinds of enterprises. In some contexts, the use of country, industry, and firm effects will help us get around the measurement challenges posed by the subjective nature of our data.

For all of their flaws, these data offer, at least in principle, one significant advantage—the directness with which they may be able to measure technology absorption. Most studies of technology transfer have tended to use firm-level productivity measures as the outcome variable. Such studies have had to contend with the reality that conventional productivity measures reflect not only the firm's mastery of more efficient methods of production, but also the ability of the firm to appropriate at least some of the benefits of that efficiency increase in the form of higher profits. However, as we have already emphasized, one can easily imagine that the entry of multinationals into local markets or trade liberalization might raise the intensity of market competition, forcing local firms to reduce markups and pass most or all of the benefits arising from increased productive efficiency on to downstream customers. Under such conditions, real productivity increases could be

---

26. Panel data (that is, data sets incorporating repeated observations on the same firms) can often, be used to determine whether a statistical relationship has a causal interpretation. When several years worth of observations are available for the same set of firms, a firm-fixed effect can be incorporated into the regression, and GMM-based regression techniques can be employed to control for the plausible endogeneity of certain key explanatory variables. These usual econometric fixes are not available to us because of the absence of a sufficiently rich panel, requiring caution to be exercised when making inferences based on regression equations such as those described above. However, we do have something of a panel dimension, in that nearly 2,000 firms appearing in the 2005 BEEPS survey also participated in the 2002 survey. We can link these firms into a short panel, and then utilize the limited panel dimension of our data to obtain some leverage around issues of causal interpretation.

27. See related work by Almeida and Fernandes (2007) for a discussion of this issue.

taking place, but their effects could be masked in the data by changes in appropriability conditions.

Going further, one could even point out circumstances in which technology absorption can create social gains, even though the cost of this acquisition of a new technical competence is so high for the acquiring firms that "productivity" benefits (that is, expansion in output that exceeds the cost of the expansion in inputs) may be slow to emerge. Technology absorption can facilitate the diversification of the economy, expand the range of diversified intermediate inputs available to local producers; raise the return to investment in sophisticated physical capital and raise the return to investment in human capital. All of these changes enhance the economic growth process, and this shows further how a narrow focus on firm-level TFP residuals can fail to fully account for the social gains generated by technology absorption, both empirically and conceptually.

The indicator variables mentioned above are more direct measures of technology transfer or technology absorption than are changes in productivity. While the measures have the disadvantage of reflecting the self-assessment of a firm's representative, they have the important advantage of not being obscured by changes in the firm's market environment that are coincident with the absorption of the new technology. Each of these outcomes could be used separately in a regression analysis; they could also be combined into an aggregate measure of technology/quality upgrading. Fortunately for us, the BEEPS data sets *also* include data on output and input measures, allowing for the computation of simple productivity measures that could also be employed in a manner more directly comparable with the previous literature.[28]

We describe below the results of regression analyses using data drawn from BEEPS surveys conducted in 2002 and 2005. Given the relationship between R&D expenditure and technology absorption emphasized by Cohen and Levinthal (1989), we believe it is important to control for R&D investment when such data are available. We will also present some results from a panel of firms that were included in both the 2002 and 2005 surveys.

## *Results from Regression Analysis*

Table 3.3 presents results from a version of equation (1) in which the dependent variable is a binary response variable indicating whether or not the firm has introduced a new product or process. We regress this on several measures of international connectedness, including: 1) a dummy variable indicating whether or not the firm has exported at all, 2) a variable measuring the fraction of sales exported, 3) a variable indicating whether the firm is majority owned by a foreign enterprise, 4) a variable measuring the fraction of sales to multinational corporations (as a proxy for participation in global supply chains), and 5) a dummy variable indicating the existence of a joint venture with a multinational. As control variables, we include the firm size, age of the firm, the log of

---

28. Preliminary empirical work confirms a positive, statistically significant connection between the discrete measures of technological absorption tracked in this study and simple measures of firm-level labor productivity.

### Table 3.3. Regression Results for New Product Introductions as Dependent Variable

Dependent Variable: New Product

| | (1) | (2) | (3) | (4) | (5) | (6) |
|---|---|---|---|---|---|---|
| Exporter dummy | 0.065<br>[0.014]*** | | | | | 0.055<br>[0.014]*** |
| Exports as a percentage of sales | | 0.001<br>[0.000]** | | | | |
| Majority foreign owned | | | 0.009<br>[0.017] | | | −0.002<br>[0.018] |
| Percentage sales to MNCs | | | | 0.001<br>[0.000] | | 0<br>[0.000] |
| Joint ventures with MNCs | | | | | 0.24<br>[0.023]*** | 0.23<br>[0.024]*** |
| Size | 0.009<br>[0.008] | 0.011<br>[0.008] | 0.013<br>[0.008]* | 0.008<br>[0.008] | 0.01<br>[0.008] | 0.002<br>[0.008] |
| Age | −0.001<br>[0.000]*** | −0.001<br>[0.000]*** | −0.001<br>[0.000]*** | −0.001<br>[0.000]*** | −0.001<br>[0.000]*** | −0.001<br>[0.000]*** |
| State ownership | −0.052<br>[0.021]** | −0.054<br>[0.021]** | −0.053<br>[0.021]** | −0.043<br>[0.022]** | −0.049<br>[0.021]** | −0.038<br>[0.021]* |
| R&D expenditures | 0.027<br>[0.004]*** | 0.029<br>[0.004]*** | 0.029<br>[0.004]*** | 0.03<br>[0.004]*** | 0.028<br>[0.004]*** | 0.027<br>[0.004]*** |
| Foreign ownership | 0<br>[0.000] | 0<br>[0.000] | | | 0<br>[0.000] | 0<br>[0.000] |
| Web use | 0.061<br>[0.012]*** | 0.064<br>[0.012]*** | 0.066<br>[0.012]*** | 0.066<br>[0.012]*** | 0.062<br>[0.012]*** | 0.058<br>[0.012]*** |
| ISO certification | 0.154<br>[0.016]*** | 0.158<br>[0.016]*** | 0.158<br>[0.016]*** | 0.155<br>[0.016]*** | 0.133<br>[0.016]*** | 0.129<br>[0.016]*** |
| Training | 0.074<br>[0.011]*** | 0.076<br>[0.011]*** | 0.076<br>[0.011]*** | 0.078<br>[0.011]*** | 0.072<br>[0.011]*** | 0.071<br>[0.011]*** |
| Skilled workforce | 0.001<br>[0.000]** | 0.001<br>[0.000]*** | 0.001<br>[0.000]*** | 0.001<br>[0.000]*** | 0.001<br>[0.000]*** | 0.001<br>[0.000]*** |
| University graduates | 0.001<br>[0.000]*** | 0.001<br>[0.000]*** | 0.001<br>[0.000]*** | 0.001<br>[0.000]*** | 0.001<br>[0.000]*** | 0.001<br>[0.000]*** |
| Infrastructure index | 0.021<br>[0.011]* | 0.02<br>[0.011]* | 0.021<br>[0.011]* | 0.019<br>[0.011]* | 0.018<br>[0.011] | 0.016<br>[0.011] |
| Governance index | 0.08<br>[0.033]** | 0.082<br>[0.033]** | 0.083<br>[0.033]** | 0.089<br>[0.033]*** | 0.084<br>[0.033]*** | 0.086<br>[0.033]*** |
| Use of a loan | 0.074<br>[0.010]*** | 0.076<br>[0.010]*** | 0.076<br>[0.010]*** | 0.077<br>[0.010]*** | 0.074<br>[0.010]*** | 0.074<br>[0.010]*** |
| Observations | 7964 | 7960 | 7986 | 7819 | 7986 | 7799 |
| R-squared | 0.16 | 0.16 | 0.16 | 0.16 | 0.17 | 0.17 |
| Country dummies | Yes | Yes | Yes | Yes | Yes | Yes |
| Industry dummies | Yes | Yes | Yes | Yes | Yes | Yes |

*Note:* Logit Regressions; Marginal effects at mean values from logit regressions shown. Dependent variable is response to question: Have you introduced a new product or process?

R&D is measured in logs in this and subsequent tables. We do not present another set of regressions that included K/L as a dependent variable, since its inclusion leads to the loss of a number of observations. Qualitatively, the results are similar to those shown below.

*Source:* Authors' Calculations, BEEPS 2002, 2005.

reported R&D expenditure, industry dummy variables, and country-of-origin dummy variables. To probe the robustness of these results and to examine the role of human capital in the absorption process, we also include controls that measure the levels of education and skill within the workforce and the existence of training programs. In most regressions, we are especially concerned about the impact of openness on local firms that are not foreign owned, and, therefore, we include an additional regressor, which measures the level of foreign ownership in the firm. We also include measures of whether firms absorb modern knowledge as those that *(i) have obtained ISO 9000 certification (a proxy for a firm's adoption of international standards and technical regulations)*, or *(ii) use the Internet for their business operations*. Finally, we also incorporate measures of infrastructure and governance into our regressions. The governance index[29] is the first principal component derived from factor analysis based on measures of the quality of contract enforcement, the predictability of government regulations, and the bureaucratic burden felt by firms. The infrastructure index[30] is the first principal component derived from the factor analysis based on infrastructure measures of power and telecom. For the most part, inclusion of this rather extensive group of additional controls did not qualitatively alter the empirical results. Where there were important changes, these will be noted explicitly in the text. Given the dichotomous nature of the dependent variable, we employed a logit specification. The nonlinear nature of the regression procedure requires us to take some care in interpreting the regression coefficients, which show the marginal effects at mean values for these regressions. The variables used in the regressions are defined in Table B.1.

Controlling for firm size, age, and ownership, one finds that larger, as well younger firms, are introducing new products and technologies. Further, state ownership decreases the likelihood of firms absorbing new technologies and processes. Our focus, however, is on the importance of various measures of international connectedness in this process of technology diffusion.

Column 1 of Table 3.3 provides the results obtained when we measure international connectedness with the export dummy variable. The coefficient is quantitatively large and statistically significant, suggesting a strong association between exporting activity and new product or process introductions. The logit regression coefficient does not have an elasticity interpretation, but evaluated at the mean values of the data, the regression coefficient implies that being an exporter, compared to being a non exporter, is associated with a higher likelihood of the introduction of new products or processes of about 6.5 percent, and this value is also highly statistically significant.

---

29. Governance Index is the first principal component based on the factor analysis of: trust in the legal system; average percentage of firms indicating that they think government regulations are predictable; (1 minus average percentage of firms indicating that it is common for firms in the industry to pay bribes to get things done); (100 minus average percentage of time managers spend dealing with regulations and inspections), by region.

30. Infrastructure Index is the first principal component based on the following variables: energy outages and telecom outages.

Column 2 gives results obtained when we measure international connectedness by employing measures of the fraction of sales that are exported. Evaluated at the means of the data, the regression coefficient implies that an increase in the export sales ratio implies an increase in the likelihood of the firm introducing a new product or process. With the full controls for investment climate and foreign ownership, the coefficient is significant at the 5 percent level. Viewed together, these first two columns appear to suggest that there is a strong association between the engagement of a firm in export activity and measures of technology absorption, but that the exact level of export activity is a weaker predictor of technology absorption.

Column 3 suggests that majority-owned foreign affiliates are more likely to introduce a new product or process, but these results are not statistically significant at conventional levels. We will later argue that this is consistent with a systematic difference in the way managers at foreign-owned firms and managers at indigenous firms interpret questions about the introduction of "new" products. It may also reflect differences in the degree to which foreign-owned firms had converged to international best practices, relative to indigenous firms, by the beginning of the current decade.

Column 5 gives the results obtained when we use a dummy variable measuring the existence of a joint venture (JV) with a multinational firm. Again, the results suggest a statistically significant positive association. Evaluated at the means of the data, the regression coefficient implies that a JV is associated with a higher likelihood of new product or process innovation of more than 100 percent.

Column 6 gives the results obtained when we include export, majority foreign ownership, sales to multinational corporations (MNCs) and JVs with MNCs, all together. The results suggest that our results are robust: The effects of all these variables remain the same as before, the only difference being that the "vertical FDI" measure is now statistically insignificant, though only marginally. Everything else is the same, which is reassuring.

We see a similar pattern when we adopt an alternative measure of technology absorption: the upgrading of an existing product or process. The results are provided in Table 3.4, which has a column format similar to that of Table 3.3.

As the reader can see, all but one of our measures of international connectedness are positively associated with technology upgrading, and each of the positive relationships is statistically significant at conventional levels, even in our most complete regression specification (except for the percentage of sales to MNCs). In keeping with the pattern established in Table 3.3, the coefficients on the export and multinational JV dummy variables are particularly large. In this table, the logit regression coefficients imply a 5 percent increase in the likelihood of product or process upgrading, as one moves from non exporter to exporter status, and a nearly 20 percent increase in the likelihood of product or process upgrading as one moves from the absence to the presence of a multinational JV.

However, the dummy variable signifying majority foreign ownership is not statistically significant at conventional levels, and this is a pattern that we see repeated in many of the regression tables that follow. Does this suggest a more limited technology transfer to wholly owned or majority owned subsidiaries? Is majority ownership by foreign firms a handicap in the technology absorption process? That interpretation would be

### Table 3.4. Regression Results for Product Upgrades as Dependent Variable

**Dependent Variable: Product Upgrade**

| | (1) | (2) | (3) | (4) | (5) | (6) |
|---|---|---|---|---|---|---|
| Exporter dummy | 0.054 [0.015]*** | | | | | 0.043 [0.016]*** |
| Exports as a percentage of sales | | 0.001 [0.000]** | | | | |
| Majority foreign owned | | | −0.021 [0.019] | | | −0.003 [0.020] |
| Percentage sales to MNCs | | | | 0.001 [0.000]* | | 0.001 [0.000] |
| Joint ventures with MNCs | | | | | 0.206 [0.025]*** | 0.195 [0.026]*** |
| Size | 0.002 [0.008] | 0.003 [0.008] | 0.003 [0.008] | −0.002 [0.008] | 0 [0.008] | −0.006 [0.008] |
| Age | 0 [0.000] | 0 [0.000] | 0 [0.000] | 0 [0.000] | 0 [0.000] | 0 [0.000] |
| State ownership | −0.1 [0.023]*** | −0.102 [0.023]*** | −0.101 [0.023]*** | −0.091 [0.023]*** | −0.098 [0.023]*** | −0.088 [0.023]*** |
| R&D expenditures | 0.025 [0.004]*** | 0.026 [0.004]*** | 0.027 [0.004]*** | 0.028 [0.004]*** | 0.026 [0.004]*** | 0.025 [0.004]*** |
| Foreign ownership | 0 [0.000] | 0 [0.000] | | 0 [0.000] | 0 [0.000] | |
| Web use | 0.1 [0.013]*** | 0.103 [0.013]*** | 0.105 [0.013]*** | 0.103 [0.013]*** | 0.101 [0.013]*** | 0.098 [0.013]*** |
| ISO certification | 0.141 [0.017]*** | 0.145 [0.017]*** | 0.146 [0.017]*** | 0.142 [0.017]*** | 0.124 [0.017]*** | 0.119 [0.017]*** |
| Training | 0.113 [0.012]*** | 0.114 [0.012]*** | 0.116 [0.012]*** | 0.113 [0.012]*** | 0.112 [0.012]*** | 0.107 [0.012]*** |
| Skilled workforce | 0.001 [0.000]** | 0.001 [0.000]*** | 0.001 [0.000]*** | 0.001 [0.000]*** | 0.001 [0.000]*** | 0.001 [0.000]*** |
| University graduates | 0 [0.000] | 0 [0.000] | 0 [0.000] | 0 [0.000] | 0 [0.000] | 0 [0.000] |
| Infrastructure index | −0.015 [0.012] | −0.015 [0.012] | −0.014 [0.012] | −0.014 [0.012] | −0.017 [0.012] | −0.018 [0.012] |
| Governance index | 0.153 [0.035]*** | 0.155 [0.035]*** | 0.155 [0.035]*** | 0.157 [0.036]*** | 0.156 [0.035]*** | 0.156 [0.036]*** |
| Use of a loan | 0.07 [0.011]*** | 0.072 [0.011]*** | 0.072 [0.011]*** | 0.074 [0.011]*** | 0.071 [0.011]*** | 0.071 [0.011]*** |
| Observations | 7964 | 7960 | 7986 | 7819 | 7986 | 7799 |
| R-squared | 0.17 | 0.17 | 0.17 | 0.16 | 0.17 | 0.17 |
| Country dummies | Yes | Yes | Yes | Yes | Yes | Yes |
| Industry dummies | Yes | Yes | Yes | Yes | Yes | Yes |

Standard errors in brackets
* significant at 10%; ** significant at 5%; *** significant at 1%.
*Notes:* Logit Regressions; Dependent variable is response to question: Have you upgraded an existing product line/service? Marginal effects at mean values from logit regressions shown.
*Source:* Authors' Calculations, BEEPS 2002, 2005.

inconsistent with one of the most well-established empirical regularities in the micro TFP spillovers literature: the finding that wholly or partially owned foreign firms tend to be more productive than other firms in the industry. In fact, when we construct simple measures of labor productivity using the BEEPS data, we find similar evidence of a positive relationship between being majority owned by foreign firms and having higher productivity.

We interpret these results in a different way: affiliates of multinationals are likely to acquire a high *level* of technology soon after acquisition; hence, they are less likely to have to upgrade. This is consistent with the studies that find that foreign affiliates are more productive—that they have a higher *level* of technology and productivity.[31]

This particular aspect of the general pattern of our results should be quite reassuring for those who might be concerned that the benefits of trade and FDI openness are largely restricted to enterprises that are controlled by foreign investors. We seem to be finding just the opposite—majority foreign ownership does *not* appear to be strongly associated with our measures of technology absorption. Rather, it appears that firms controlled by *local* owners, engaged in exports, or participants in FDI-mediated supplier networks are the primary beneficiaries.

A further alternative perspective on this phenomenon is provided by Table 3.5, which measures whether the firm in question has acquired a new production technology.

The pattern of regression coefficients is similar to that in the previous tables: All measures of international connectedness, except majority foreign ownership, are found to be positively correlated with the acquisition of new technology. All of the positive correlations are statistically significant at the 10 percent level, even in the most complete specifications, and the impact of exports and multinational JVs is especially large and significant at the 1 percent level.

A final examination of the correlation in the cross-section comes from Table 3.6. Here, we construct a dependent variable, "technology upgrading," from a composite of dummy variables measuring firms' introduction of new products or processes, upgrading of existing products or processes, achievement of new quality certifications, and acquisition of technology licensing agreements from other firms. We simply sum up responses by firm, producing a limited dependent variable that lies between 0 and 4. We apply ordinary least squares regression analysis, controlling for country and industry fixed effects. The results of these regressions parallel our earlier results. Once again, all measures of international connectedness, except majority foreign ownership, are positively and significantly associated

---

31. An alternative interpretation of this result is that majority foreign-owned firms may apply a different set of criteria in determining whether a particular technology is "new," and in determining whether a particular process or product is being upgraded. It is quite likely that the managers of foreign firms will either be foreigners themselves, or they will be indigenous managers with a very deep understanding of technological best practices as it exists in the global economy beyond the ECA region. Such individuals are less likely to flag a process or technology as "new" when it only brings the ECA enterprise up to standard practice elsewhere, whereas an indigenous enterprise manager might be much more likely to view adoption of the identical process or technology as "new." To the extent that our interpretation is correct, it suggests that empirical results regarding the impact of majority foreign ownership have to be viewed with care and a certain degree of skepticism.

## Table 3.5. Regression Results for Introduction of New Technology as Dependent Variable

| Dependent Variable: New Technology | | | | | | |
|---|---|---|---|---|---|---|
| Exporter dummy | 0.053 [0.014]*** | | | | | 0.051 [0.015]*** |
| Exports as a percentage of sales | | 0.001 [0.000]* | | | | |
| Majority foreign owned | | | −0.065 [0.018]*** | | | −0.076 [0.018]*** |
| Percentage sales to MNCs | | | | 0 [0.000] | | 0 [0.000] |
| Joint ventures with MNCs | | | | | 0.088 [0.024]*** | 0.082 [0.024]*** |
| Size | 0.016 [0.008]** | 0.017 [0.008]** | 0.019 [0.008]** | 0.015 [0.008]** | 0.017 [0.008]** | 0.012 [0.008] |
| Age | −0.001 [0.000]*** | −0.001 [0.000]*** | −0.001 [0.000]*** | −0.001 [0.000]*** | −0.001 [0.000]*** | −0.001 [0.000]*** |
| State ownership | −0.039 [0.021]* | −0.04 [0.021]* | −0.04 [0.021]* | −0.037 [0.022]* | −0.038 [0.021]* | −0.037 [0.022]* |
| R&D expenditures | 0.032 [0.004]*** | 0.033 [0.004]*** | 0.034 [0.004]*** | 0.034 [0.004]*** | 0.034 [0.004]*** | 0.032 [0.004]*** |
| Foreign ownership | −0.001 [0.000]*** | −0.001 [0.000]*** | | −0.001 [0.000]*** | −0.001 [0.000]*** | |
| Web use | 0.068 [0.012]*** | 0.071 [0.012]*** | 0.073 [0.012]*** | 0.073 [0.012]*** | 0.071 [0.012]*** | 0.068 [0.012]*** |
| ISO certification | 0.141 [0.016]*** | 0.143 [0.016]*** | 0.144 [0.016]*** | 0.145 [0.016]*** | 0.135 [0.016]*** | 0.133 [0.016]*** |
| Training | 0.088 [0.011]*** | 0.089 [0.011]*** | 0.09 [0.011]*** | 0.09 [0.011]*** | 0.088 [0.011]*** | 0.086 [0.011]*** |
| Skilled workforce | 0 [0.000] | 0 [0.000] | 0 [0.000] | 0 [0.000] | 0 [0.000] | 0 [0.000] |
| University graduates | 0 [0.000] | 0 [0.000] | 0 [0.000] | 0 [0.000] | 0 [0.000] | 0 [0.000] |
| Infrastructure index | −0.023 [0.011]** | −0.023 [0.011]** | −0.023 [0.011]** | −0.022 [0.011]** | −0.024 [0.011]** | −0.024 [0.011]** |
| Governance index | 0.007 [0.033] | 0.01 [0.033] | 0.011 [0.033] | 0.009 [0.033] | 0.011 [0.033] | 0.006 [0.033] |
| Use of a loan | 0.06 [0.010]*** | 0.062 [0.010]*** | 0.062 [0.010]*** | 0.063 [0.010]*** | 0.062 [0.010]*** | 0.061 [0.010]*** |
| Observations | 7901 | 7897 | 7920 | 7754 | 7920 | 7736 |
| R-squared | 0.15 | 0.15 | 0.15 | 0.15 | 0.15 | 0.15 |
| Country dummies | Yes | Yes | Yes | Yes | Yes | Yes |
| Industry dummies | Yes | Yes | Yes | Yes | Yes | Yes |

Standard errors in brackets
* significant at 10%; ** significant at 5%; *** significant at 1%.
*Note:* Logit regressions with dependent variable being response to question: Have you acquired a new production technology?
*Source:* Authors' Calculations, BEEPS 2002, 2005.

with technology upgrading, as measured by this composite dependent variable. The statistical significance holds at the 1 percent level for all positive relationships, even in the most complete specifications.

Another empirical regularity that merits comment is the robustly positive relationship that exists between measures of human capital at the firm level and technology absorption. We incorporate three variables to measure human capital: a dummy variable indicating the existence of an in-house training program; a ratio of the number of professionals to total employees; and a ratio of the number of university graduates to total employees. The inherent ambiguity of the definition of a "professional," and imperfect information on the part of the survey respondent concerning the educational attainment of employees, likely leads to significant measurement error in the latter two variables. Nevertheless, in nearly all regressions, the estimated coefficients for at least two of these three variables are positive and statistically significant. The results for training suggest particularly strong relationships between this variable and our outcome measures. Evaluated at the mean of the data, the coefficients for training imply that the existence of a program raises the probability of technology absorption by 40 percent or more.

The regression analysis also controls for the impact of information and communication technology (ICT) and quality standards on technology absorption. ICT is considered the preeminent "general purpose technology" of the past 20 years, as it has driven economy wide growth over a range of sectors by prompting them to innovate and upgrade further, with technological progress in these sectors in turn creating incentives for further advances in the ICT sector, thus setting up a positive, self-sustained virtuous cycle (Bresnahan and Trajtenberg 1995; Helpman and Trajtenberg 1996). Similarly, there is strong evidence across a range of sectors that the adoption of industry standards and technical regulations are among the most important forms of introducing product and process technology upgrades and increasing productivity for firms (Corbett, Montes-Sancho, and Kirsch 2005). Standards thus play a particularly central role in diffusing knowledge in those industries where products and processes supplied by various providers must interact with one another.

The regression results show that both Internet use and ISO certification are positively and significantly related at the 1 percent level to all the measures of technology absorption. Depending on its international connectedness, a firm in ECA that uses the Internet is between 9.8 percent and 10.5 percent more likely to have upgraded a product or process, and between 5.8 percent and 6.6 percent more likely to have introduced a new product. The coefficients for ISO certification suggest that a firm that is ISO certified is between 11.9 percent and 14.6 percent more likely to have undertaken the upgrading of a product or process, and the probability of it having introduced a new product is between 12.9 percent and 15.8 percent higher. Further, a firm that uses the Internet is between 6.8 percent and 7.3 percent more likely to have purchased new technology in the last three years, while an ISO-certified firm is between 13.3 percent and 14.5 percent more likely to have done so than a firm that isn't certified. Finally, the likelihood of a firm in ECA having upgraded its technology, as measured by this composite dependent variable, increases by between 30 percent and 34 percent if the firm uses the Internet. Appendix D analyzes the determinants of ISO certification and Internet use by firms in ECA countries using the 2002 and 2005 World Bank BEEPS. We outline a framework that highlights three types of

## Table 3.6. Linear Regression Results Based on a "Composite" Measure of Absorption

**Dependent Variable: Composite Measure "Upgrade"**

| | (1) | (2) | (3) | (4) | (5) | (6) |
|---|---|---|---|---|---|---|
| Exporter dummy | 0.252 [0.038]*** | | | | | 0.186 [0.038]*** |
| Exports as a percentage of sales | | 0.002 [0.001]*** | | | | |
| Majority foreign owned | | | −0.084 [0.048]* | | | −0.129 [0.048]*** |
| Percentage sales to MNCs | | | | 0.004 [0.001]*** | | 0.003 [0.001]*** |
| Joint ventures with MNCs | | | | | 1.137 [0.062]*** | 1.099 [0.064]*** |
| Size | 0.091 [0.021]*** | 0.099 [0.021]*** | 0.105 [0.021]*** | 0.091 [0.021]*** | 0.083 [0.021]*** | 0.065 [0.021]*** |
| Age | −0.003 [0.001]*** | −0.002 [0.001]** | −0.002 [0.001]** | −0.002 [0.001]** | −0.002 [0.001]** | −0.002 [0.001]*** |
| State ownership | −0.188 [0.058]*** | −0.194 [0.058]*** | −0.195 [0.058]*** | −0.167 [0.059]*** | −0.177 [0.057]*** | −0.154 [0.058]*** |
| R&D expenditures | 0.15 [0.011]*** | 0.158 [0.011]*** | 0.16 [0.011]*** | 0.16 [0.011]*** | 0.15 [0.011]*** | 0.143 [0.011]*** |
| Foreign ownership | −0.001 [0.001]** | −0.001 [0.001]* | | −0.001 [0.001] | −0.001 [0.001]* | |
| Web use | 0.32 [0.032]*** | 0.338 [0.032]*** | 0.344 [0.032]*** | 0.338 [0.032]*** | 0.316 [0.031]*** | 0.301 [0.032]*** |
| Training | 0.46 [0.029]*** | 0.468 [0.029]*** | 0.472 [0.029]*** | 0.469 [0.030]*** | 0.441 [0.029]*** | 0.431 [0.029]*** |
| Skilled workforce | 0.003 [0.001]*** | 0.004 [0.001]*** | 0.004 [0.001]*** | 0.004 [0.001]*** | 0.003 [0.001]*** | 0.003 [0.001]*** |
| University graduates | 0.001 [0.001]* | 0.001 [0.001]** | 0.001 [0.001]** | 0.001 [0.001]** | 0.001 [0.001]* | 0.001 [0.001] |
| Infrastructure index | −0.016 [0.030] | −0.016 [0.030] | −0.016 [0.030] | −0.021 [0.031] | −0.03 [0.030] | −0.035 [0.030] |
| Governance index | 0.381 [0.089]*** | 0.391 [0.089]*** | 0.393 [0.089]*** | 0.41 [0.090]*** | 0.392 [0.087]*** | 0.396 [0.088]*** |
| Use of a loan | 0.246 [0.028]*** | 0.256 [0.028]*** | 0.255 [0.027]*** | 0.256 [0.028]*** | 0.245 [0.027]*** | 0.24 [0.027]*** |
| Observations | 7901 | 7897 | 7920 | 7754 | 7920 | 7736 |
| R-squared | 0.26 | 0.25 | 0.25 | 0.26 | 0.28 | 0.29 |
| Country dummies | Yes | Yes | Yes | Yes | Yes | Yes |
| Industry dummies | Yes | Yes | Yes | Yes | Yes | Yes |

Standard errors in brackets
* significant at 10%; ** significant at 5%; *** significant at 1%.

*Note:* The dependent variable "upgrade" is the sum of firm responses to questions on: 1) introduction of a new product or process, 2) upgrading of existing product or process, 3) achievement of new quality certification, and 4) new technology licensing agreement. We ran an ordered logit regression for the same specification and got the same results qualitatively confirming the robustness of the relationship.
*Source:* Authors' Calculations, BEEPS 2002, 2005.

factors—input markets, market incentives, and access to international knowledge-that may be associated with ISO certification and Internet use.[32]

We also incorporate measures of infrastructure and institutional quality into our regressions in Tables 3.3 to 3.6. While we think it is important to incorporate these variables as controls, it is difficult to come to general conclusions regarding the estimated coefficients on these variables. Access to finance, which is proxied by a firm's use of a loan,[33] has a strong positive and statistically significant relation to the technology absorption variables. This would seem intuitive, since financially constrained firms would find it harder to invest in technology upgrading processes and methods.[34] The governance index that is a combination of measures of corruption, trust in the legal system and institutional regime, has a positive effect on new product innovations and upgrades, as well as on the composite upgrade variable. It does not, however, have an effect on the introduction of a new technology. The infrastructure index that captures the quality of the communications network is only positively associated with the introduction of a new product. This would seem to suggest that some of the impact of this variable may, in fact, be already captured by one of the other control variables.

More generally, we view the robustness of these results as encouraging. The consistency in the pattern of results that we observe across these four sets of regressions suggests that the underlying economic relationships are real. But are they causal in nature? Within the confines of these data, it is difficult to address this question definitively.[35] Nevertheless, we can exploit the panel structure of our data to obtain some degree of clarity around these

---

32. Specifically, we find that firms that have the appropriate *complementary inputs*, namely managerial capacity, higher shares of skilled labor, access to finance (and to a lesser extent access to good infrastructure), and *access to international knowledge*, either from foreign investors or by exporting, *are more likely to be ISO certified and use the Internet*. Our results suggest that the relationship between market incentives and Internet use or ISO certification is more nuanced. While pressures from consumers generate demand for ECA firms to upgrade their technology, pressures from competitors do not, which may seem counterintuitive in a developed economy setting, but is consistent with previous literature that argues that most firms in ECA face very substantial resource constraints (particularly financial resources), and thus only those with rents are able to finance activities related to knowledge absorption. Accordingly, we find that large and medium-sized firms are significantly more likely to be ISO certified, and particularly, to use the Internet. Our findings also suggest that privatized firms exhibit better ISO certification and Internet use only when there is a clear private owner with a profit incentive, who is empowered to make changes. Together with the above-mentioned finding that in environments with severe credit constraints the firms with substantial rents are more likely to absorb knowledge, this underscores the importance of improving market *incentives in conjunction with efforts to* improve access to knowledge and to critical complementary inputs.

33. The use of a loan is a dummy = 1 if the firm has a loan.

34. We also estimated the equation with access to finance being proxied for by "the perception of firms that feel access to finance was a major obstacle." Using this variable, however, produced counterintuitive results, with firms that expressed a greater difficulty in accessing financial services showing a greater probability of absorbing new technologies and introducing new products.

35. The unavailability of a reliable IV made it difficult to implement the instrumental variables approach to deal with the issue of endogeneity. Attempts were made in the direction, but were unsuccessful. We experimented with the location of firms as an exogenous and predetermined indicator of international connectedness, reasoning that the locations of firms in business prior to the transition were not determined by market forces. On the other hand, firms more proximate to Western markets might be more likely to establish relatively higher degrees of international connectedness. These attempts did not yield informative results.

issues. Table 3.7 presents the results of a panel regression in which we present results based on the roughly 1,700 firms that were surveyed in BEEPS in both 2002 and 2005.

We run a fixed-effects regression on these data. A natural concern in this context is that firms with "better" management will seek both international connectedness and technology

### Table 3.7. Regression Results Based on Composite Measure of Absorption, Panel Data

**Dependent Variable: Composite Measure "Upgrade"**

| | (1) | (2) | (3) | (4) | (5) |
|---|---|---|---|---|---|
| Exporter dummy | 0.388 [0.115]*** | | | | 0.274 [0.120]** |
| Exports as a percentage of sales | | 0.004 [0.002]* | | | |
| Percentage sales to MNCs | | | 0 [0.003] | | −0.001 [0.003] |
| Joint ventures with MNCs | | | | 0.392 [0.130]*** | 0.372 [0.137]*** |
| Size | 0.156 [0.068]** | 0.164 [0.068]** | 0.17 [0.072]** | 0.173 [0.068]** | 0.158 [0.071]** |
| Age | −0.001 [0.007] | −0.001 [0.007] | 0 [0.007] | −0.001 [0.006] | −0.001 [0.007] |
| State ownership | −0.618 [0.244]** | −0.596 [0.243]** | −0.518 [0.252]** | −0.6 [0.244]** | −0.52 [0.250]** |
| Foreign ownership | −0.002 [0.002] | −0.001 [0.002] | −0.002 [0.002] | −0.002 [0.002] | |
| Web use | 0.108 [0.084] | 0.129 [0.085] | 0.118 [0.086] | 0.1 [0.084] | 0.121 [0.086] |
| Training | 0.253 [0.070]*** | 0.26 [0.070]*** | 0.263 [0.071]*** | 0.257 [0.070]*** | 0.277 [0.071]*** |
| Skilled workforce | 0.001 [0.002] | 0.001 [0.002] | 0.001 [0.002] | 0.001 [0.002] | 0 [0.002] |
| Infrastructure index | 0.146 [0.138] | 0.136 [0.138] | 0.083 [0.141] | 0.138 [0.138] | 0.113 [0.140] |
| Governance index | 0.125 [0.185] | 0.126 [0.185] | 0.073 [0.187] | 0.115 [0.185] | 0.025 [0.187] |
| Use of a loan | 0.112 [0.080] | 0.091 [0.081] | 0.094 [0.081] | 0.096 [0.080] | 0.113 [0.082] |
| Observations | 2366 | 2355 | 2308 | 2377 | 2296 |
| Firm fixed effects | Yes | Yes | Yes | Yes | Yes |
| R-squared | 0.04 | 0.03 | 0.03 | 0.04 | 0.04 |

Standard errors in brackets
* significant at 10%; ** significant at 5%; *** significant at 1%

*Notes:* Linear regressions based on "composite" measure of absorption, as above. Dependent variable is the sum of firm responses to four questions: 1) Have you introduced a new product or process? 2) Have you upgraded an existing product or process? 3) Have you acquired a new quality certification? 4) Have you entered into a new product or technology licensing agreement?
*Source:* Authors' Calculations, BEEPS 2002, 2005.

absorption, and that the positive associations in the cross-section merely reflect the mutual impact of superior management on both of these variables. By running a fixed-effects regression, we effectively extract the impact of all firm-specific variables that do not vary over time. One might logically think that superior management would be exactly this kind of variable, assuming that the management did not change in the firm over the period analyzed. The cost of this fixed-effects approach is that the cross-sectional dimension of variance is largely extracted from the data; one is left with inference based on changes within firms over time. If variables are measured with error, the fixed-effects approach can significantly worsen the measurement bias, leading to an attenuation of measured effects (Hausman and Griliches 1986). Measurement error is likely to be particularly severe in data sets like ours. In addition, two key variables-the level of R&D spending reported by the firm, and the fraction of employees who are university graduates—are not available in the earlier survey year. We are thus not able to estimate coefficients for these variables. Nevertheless, we proceed to regress our composite measure of technology upgrading on our five measures of international connectedness, using a linear fixed-effects approach.

Larger firms continue to be more likely to absorb new technologies. The coefficient on the dummy variable indicating majority foreign ownership could not be estimated, because almost no firms in our sample changed their majority foreign ownership status between the 2002 and 2005 sample years. As a consequence, we omit this variable from consideration. However, the coefficients on our export and multinational JV dummies continue to indicate economically and statistically significant associations between these measures of connection to the global economy and to technology absorption. The coefficients on the fraction of sales going to multinationals are no longer significant at conventional levels. In this fixed-effects context, our coefficient estimates are driven entirely by firms that *change* their exporting status (or multinational JV status) over time. Firms that move from non-exporting to exporting, achieve an expansion in our composite technology upgrading measure of 39 percent. Firms that move from no JV to a multinational JV achieve gains of nearly 39 percent. While we cannot rule out other interpretations, these results lend themselves much more readily to a causal, or at least a partly causal, interpretation. The training coefficient is also positive and statistically significant at conventional levels. The coefficient implies that the transition from no training to a training program increases technology absorption by 25 percent to 28 percent.

## Discussion of Results with Implications for Policy

### Exporting Leads to Technological Improvements among ECA Firms

We believe that these results are useful, in part, because of the light that they cast on the debate over the existence of a "learning-by-exporting" effect. A long, well-cited series of case studies has documented the importance of the process by which firms in East Asia learned to improve their manufactured products and manufacturing processes through their efforts to export to more advanced foreign markets (Pack and Westphal 1986). However, most attempts to identify positive TFP growth effects from learning by exporting in firm-level or plant-level data, have rejected the hypothesis that a transition of firms into exporting is associated with an increase in TFP growth. Researchers have effectively concluded that learning-by-exporting effects do not exist (Bernard and Jensen 1997; Clerides, Lach, and Tybout 1996). In an ECA context, Commander and Svejnar (2007) have suggested that, controlling for foreign ownership, there is no independent effect of exporting on firm productivity.

Our results suggest a different conclusion, and one that may apply far beyond the ECA context. In our panel data, *transition* to exporting is positively and significantly correlated with *increases* in measured technology upgrading. We find this to be true even when we control for foreign ownership, human capital, and environmental factors affecting export climate. This is consistent with the hypothesis that exposure to foreign markets fosters learning, and our results suggest that this learning effect is not limited to foreign-owned firms. It is, of course, not inconsistent with the view that firms, as they seek to transition to exporting, will invest in upgrading their technology to make themselves more competitive in foreign markets. In other words, technology upgrading could also increase exports—but, to the extent that the technology upgrading was motivated in the first instance by the desire to compete in a foreign market, it still reinforces the policy implications we stress.

This finding could be reconciled with the absence of TFP growth effects in a number of ways. First, it is likely that the implementation of the learning obtained through export experience does not come for free—upgrading products and processes requires expenditures on foreign technology licenses, consultants, worker training, and new capital goods. Second, active competition with other new entrants (from the same developing country) into the first world export market could limit the ability of any one producer to appropriate the gains from learning by exporting. It would clearly be in the interests of the downstream customers of these suppliers to encourage such competition. In the absence of firm-specific output prices and cost measures, this increased competitive intensity could, as we argued earlier, squeeze measured TFP effects to something close to zero. Third, part of the gains from successful entry into foreign markets could accrue not to the firms, but to the (initially) small number of talented managers within the local managerial labor markets capable of managing world-class production operations. Regardless of which of these theories, if any, is correct, the essential policy implication is the same. This strengthens the case for further movement toward an open trade regime and to the elimination of policies that penalize or undermine exporters. ECA governments seeking to promote technology absorption should eliminate export disincentives and pursue a policy regime that provides appropriate support for export activity.

### FDI's Contribution to Technological Development of Local Firms

The positive relationships between local firms and multinationals, and our measures of technology absorption, suggest that vertical FDI does promote learning by local firms, and it identifies some explicit channels through which the learning occurs. ECA governments seeking to encourage technology absorption should continue to open themselves to FDI, and should critically examine informal barriers to foreign firm expansion that might impede this channel of technology diffusion.

Reduction of the remaining barriers to FDI in ECA could increase FDI and, given the positive association of absorption and FDI, facilitate absorption. For example, Russia fares worse than other countries in the region, attracting one of the lowest levels of FDI inflows. Related World Bank-supported research has pointed to key shortcomings in the Russian business environment. Many of these shortcomings are a function of government policies that limit FDI inflows and foreign firm operations, especially in the service sector (Desai and Goldberg 2008). The reduction in barriers to FDI in service sectors would allow all multinationals to obtain greater post-tax benefits on their investments, encouraging them to increase FDI to supply the Russian market. This, in turn, would lead to an increase in technology absorption, as implied by the positive association of FDI and absorption.

Removing barriers to trade in services in a particular sector is likely to lead to lower prices, improved quality, and greater variety. Efficient services would be vital intermediate inputs into the productive sector, and the telecom sector would be particularly crucial to the diffusion of knowledge. Technology transfer accompanying this service liberalization-either embodied in foreign direct investment, or disembodied, would have a stronger effect on growth.[36] Among the key restrictions on foreign service providers in Russia, for example, are the monopoly of Rostelecom on fixed-line telephone service, the prohibition of affiliate branches of foreign banks, and the restricted quota on the share of multinationals in the insurance sector. The case of Russian accession to the World Trade Organization provides substantial and robust evidence that various measures of regulation in the product market, particularly entry barriers, are negatively related to investment. The implications of our analyses are clear: regulatory reforms, especially those that liberalize entry, are very likely to spur investment (Jensen, Shepotylo, and Tarr 2007; Alesina and others 2005).

Another example of the remaining reform agenda in ECA is Kazakhstan, which has done more to lower its tariffs on goods than it has to liberalize its barriers to FDI in the service sectors (Jensen and Tarr 2007). In telecommunications, there is strong restriction on entry in some areas of service; incumbent operators have excessively long exclusivity rights in areas such as long-distance and international telephone services, and with respect to interconnection for mobile operators; foreign ownership of a company cannot exceed 49 percent; and competition is limited due to cross-ownership among the incumbents. The costs of long-distance telephone services and broadband Internet access are three to six times the costs of comparator countries such as Russia, selected EU countries, and Australia. In banking, up to the end of 2005, branches of foreign banks were prohibited, and banks with foreign participation were limited to a maximum of 50 percent of the aggregate authorized capital of the sector. In transportation services, railroad tariffs differentiate among import, export, and domestic freight destinations. Existing railroad service is quite slow, contributing to very long delivery times for external trade. In air transportation services, domestic route licensing is rather restrictive. These factors, in turn, would explain why Kazakhstan ranks low in the Logistics Performance Index.

---

36. Mattoo (2005) argues that since many services are inputs into production, the inefficient supply of such services acts as a tax on production, and prevents the realization of significant gains in productivity. As countries reduce tariffs and other barriers to trade, effective rates of protection for manufacturing industries may become negative if they continue to be confronted with input prices that are higher than they would be if services markets were competitive, making it imperative to have policies to liberalize trade in services and attract FDI in key service sectors like telecommunications and financial services.

Although technology is making it easier to trade in services, often FDI plays a vital role in selling services (Eschenbach and Hoekman 2005). Given the lack of a service sector under central planning, FDI can be expected to play a particularly important role, more so than in countries where incumbent competition confronts foreign providers. Overall, services account for some 62 percent of the stock of FDI in 12 selected ECA countries, with finance, transport, communications, and distribution services accounting for the largest share of this FDI. While the share of the service sector in GDP, employment, output per worker, trade, and FDI in Central and Eastern European countries shows substantial convergence toward that of Western European countries, it also shows a distinct difference between Central European/Baltic states and Central Asian and CIS economies. Reforms in policies regarding financial and infrastructure services, including telecommunications, power, and transport, are highly correlated with inward FDI (Eschenbach and Hoekman 2005).

Firm-level investments in human capital and the skill level of the workforce are strongly associated with technology absorption. In every regression, the presence of a worker training program was strongly associated with technology absorption, and measures of the skill level of the labor force often had significant effects on absorption. In the panel regressions, the introduction of a training program is positively associated with increases in technology absorption. The Serbia case studies, discussed in the next chapter, also highlight the importance of worker training in cases of successful technology absorption.

The relationship between training and skills on the one hand, and successful technology absorption on the other, is complex, with causality almost surely running in both directions. Training and knowledge absorption are complementary, in the sense that a firm's capacity to absorb new knowledge, and to benefit from absorption, depends on the skills and training of the workforce. Higher levels of training and skills typically lead to a firm identifying new technologies that need to be mastered in order to increase competitiveness. Yet the decision of the firm to acquire a certain technical competency often necessitates training and changes in the skill composition of the workforce. For example, training in Russian enterprises is also highly correlated with indicators of innovativeness-such as R&D or licensing of patents and know-how, introduction of new production technologies, and high technology exports (Tan and others 2007).

These findings have implications for both firm strategy and public policy in the ECA region. While all countries struggle to align the output of their formal public educational systems with the changing needs of their industries, the challenge has been particularly acute in the ECA region. The legacy of socialism included a number of significant educational achievements, but many features of the pre-reform system were not well-suited to the needs of an open, competitive economy. Despite the substantial progress that has occurred since transition, more work remains to be done. Again, in Russia, in spite of the high and rising demand for educated and skilled workers, there exist skills shortages in enterprises. The reasons for this shortage include an educational and training system that is under funded below the tertiary level and that faces numerous challenges; an industrial sector with high labor turnover (which inhibits training); and the inability of some non-competitive firms to pay competitive wages to attract and retain needed skills.

The issue of worker training also deserves consideration. In an economic environment with labor mobility, firms may be reluctant to invest in the skills of workers who might simply leave the firm and take those skills to a rival for slightly more pay. This is especially the case when firms face financial or other constraints that may limit their ability to engage in other necessary investments. While full consideration of this issue would take us beyond the scope of this study, one may be able to make a case for public-private co-investment in worker training. In essence, governments subsidize worker training in firms, but firms will always bear part of the cost themselves, ensuring that government resources are generally directed to training programs that bring real benefits. With a view to remedy this underinvestment in training, Tan and others (2007), suggest that the Russian government should consider putting in place employer-targeted training policies. ECA countries can learn from drawing on the experiences[37] of many other countries, both industrial and developing, that have used payroll-levy training funds, tax incentives for employer-sponsored training, and matching grants. They suggest that policies should be designed to increase competition in training provision from all providers, both public and private, including the employer. Further, they also cite the use of matching grants, which can help to develop a training culture. The most successful schemes are demand driven, implemented by the private sector, and intended to sustain the markets for training services. With a view to generate training capacity in enterprises and increase the propensity for workers to undertake training, grants in ECA should aim at strengthening and diversifying the supply of training and stimulating demand.

The cost is only one barrier to effective worker training. Indigenous firms behind the technology frontier are often not knowledgeable about the kinds of training programs that could effectively equip their workers to manage new technology in an efficient manner. The Serbia case studies describe intensive training efforts conducted by foreign owners to bring the acquired firms to the technical frontier. In some cases, this included bringing assembly line workers and a shop foreman into established plants in other countries, so that front-line workers could receive direct advice and instruction from their peers in the parent company. Training manuals and training procedures used in contexts like this can often be considered strategic assets of a foreign firm; there may be a natural reluctance to share such knowledge with unaffiliated indigenous firms.

Yet, there are other circumstances in which such knowledge sharing may be in the mutual interest of local and foreign firms. As documented in East Asia, foreign buyers are often willing to share detailed technical knowledge with local suppliers, enhancing the worker training process. Foreign manufacturing firms located in ECA have strong incentives to ensure that direct and indirect suppliers meet quality and efficiency standards, and there will be incentives for knowledge sharing in those contexts as well. ECA governments and firms should make the most of the opportunities that this confluence of interests creates. And, of course, this line of thinking again underscores how important it is for the region's countries to continue to embrace trade and FDI openness.

---

37. Such as those cited with regard to training levies in Middleton, Ziderman, and Adams (1993) and Gill, Fluitman, and Dar (2000).

# CHAPTER 4

# How Does FDI via Company Acquisition Impact Technology Absorption? A Case Study of Serbian Enterprises

*By John Gabriel Goddard, Itzhak Goldberg, and Wolfgang Rigler*

The objective of this case study is to complement the findings from the econometric analysis of the BEEPS and patent surveys presented in the previous chapters. Those analyses provide evidence that multinationals contribute to *indigenous* technological improvement decisions and have played an increasing role in regional patenting activity, and the case study methods are used in this chapter to provide a richer dynamic perspective of the causal relations between FDI and absorption. The need to supplement econometric findings in this area of research has been raised, *inter alia*, by Howard Pack in his survey of econometric versus case study approaches to technology transfer.[38]

Focusing on the dimensions of product mix, production technology, management, and skills, the case study sheds light on the firm-level absorption process taking place within an acquisition FDI context in a transition and post-conflict country, namely, Serbia. The case study illustrates the critical role of foreign strategic investors to help acquired companies cope with multiple challenges of absorbing knowledge, whether this is embedded in capital goods, derived from learning through exporting, comes from consultants or other knowledge brokers, or is codified in intellectual property that requires licensing. The high financial and non-pecuniary costs of absorption were stressed in the literature (see Cohen and Levinthal 1990, and more recently, Keller 2004); our in-depth

---

[38]. He argues that "econometric and case studies are complementary" because while most econometric studies of technology transfer rely on censuses or surveys that do not permit analysis of the determinants of a firm's evolution, it is difficult to generate a sample of case studies that is sufficiently large enough so the results can be viewed as robust (Hoekman and Javocik 2006).

inspection of the transition in eight manufacturing companies illustrates this point, and compares how foreign strategic investors, local investors, and insider-owners[39] confront and minimize the costs of absorption.

Several reasons explain why the Serbian government believed that attracting reputable international investors to its 2001 post-Milosevic privatization program was not only necessary, but also feasible, in spite of Serbia's post-war difficulties. As a brief historical background: in the 1990s, armed conflicts and the dissolution of the former Yugoslavia led to international sanctions, which interrupted the production relations of Yugoslav companies, which were well-established in Europe, and caused international isolation of these companies from input sources, as well as the loss of markets. These events left a dire economic legacy, and it was clear that companies (including especially 86 companies selected by for tender privatization due to their size, importance, etc.) required FDI to allow them to regain their position in European markets and to replace the technologies that had by then become obsolete. The privatization program was able to sell these companies, and this was the core of its success. Our interest in this study is in the direct effects of FDI acquisition on companies that received investment, and not—as in the spillovers literature—on other firms (competitors or suppliers, horizontally or vertically), or on the economy as a whole.

The chapter makes a distinction between "Greenfield"[40] FDI and FDI based on acquisition of existing assets from the government (privatization), or from private owners. The importance of this distinction is that countries that are poor in natural resources (like Serbia), and whose investment climates are still uncompetitive, have a hard time attracting Greenfield FDI, Therefore, selling existing assets, whether they be private, socially owned or state-owned is a more realistic option for accelerating industrial development. However, the experience of the companies we examine shows that FDI via acquisition is often a first step for making Greenfield investments.

Our emphasis on the outcomes of acquisitions of productive assets in Eastern Europe is still relevant today, because the legacy of mass privatization has frequently resulted in insider control that prevents openness to change and hampers absorption and innovation. In the policy implications chapter, we argue that governments could facilitate FDI via a properly regulated mergers and acquisition (M&A) process, if they instituted a process of consolidation of the post-privatization ownership structure in insider-dominated companies.

The chapter's next section reviews what we know from previous studies about the incentives for foreign investors to enter ECA markets through acquisition vis-à-vis Greenfield investments, and the related incentives to increase technological competitiveness by introducing new governance arrangements, product lines, managerial and workforce skills, and so on. We then explain the context for the case study and present its main results. Although we have tried to structure the case study so as to gain meaningful and hopefully robust insights about the "black box" of absorption via FDI by relying on a combination of

---

39. A "strategic foreign investor" is one that operates a business in the same industry as the acquired firm and is purchasing the business assets with the intention of operating them. In a transition context, "insider-owners" refer to managers who bought ownership stakes in former socially owned and state-owned companies.

40. Greenfield investment is a *de novo* investment in a previously undeveloped site.

qualitative interviews and financial analysis, it is important to provide an external quantitative check. For this purpose, we draw on the results of the 2005 EU impact study of the Serbian privatization.[41] The final sections discuss the lessons for policy and conclusions.

The following questions guide the research and discussion in this chapter:

1. What is the effect of foreign ownership on technology absorption—that is, is there a difference in the absorption process followed by firms acquired by a local *versus* a foreign investor such as a multinational enterprise?
2. How does ownership affect corporate governance, and how does the latter, in turn, affect absorption?
3. Was management and organizational change a prerequisite for the implementation of new investments and technology?
4. What were the effects of the investment climate on M&A FDI, and what are the corollary effects for Greenfield FDI?
5. What determined the foreign investors' relocation of R&D to and from Serbia?

## Investment Climate and Sequence of Mergers and Acqusitions, and Greenfield FDI

Most FDI in Serbia in the period 2001–06 was related to privatization. The total stock of FDI was US$8.9 billion at the end of 2006, of which US$4.3 billion was in Greenfield investments, mainly in retail. However, a substantial share of this Greenfield investment is actually after-privatization investment by the new owners, in order to improve and scale up existing operations and ensure the quality of products. For instance, the Danish company Carlsberg bought a brewery in the town of Celarevo for US$5 million, but then invested an additional US$20 million in order to reach its required production quality. The additional US$20 million investment is not captured in the statistics as privatization revenue, but as Greenfield investment.

We will show that acquisition (also called Brownfield) investments in Serbia play a positive role in encouraging Greenfield investors, partly because new owners encourage investments by suppliers and contractors that they have relationships in other countries where they operate. Needless to say, this requires that current investors find the investment climate acceptable. In recent years, Serbia has seen concrete improvements in its investment climate (see World Bank *Doing Business 2006*, and for background, Goldberg, Radulovic, and Schaffer 2005).

In addition to the domestic investment climate, the potential for attracting FDI will depend on regional and global variables that are not under the control of governments. Economies of scale are very strong in industries facing high fixed setup costs, and become a critical factor shaping the global investment allocation of multinationals. A recent IMF paper by Damekas and others (2005) claims that: "Virtually all empirical studies find that gravity

---

41. The Impact Study is based on a survey of 187 companies, of which 122 were privatized under the 2001 law and 65 under the 1997 law. The Impact Study focuses on the differences between the two groups. Three of our case studies were privatized under the 2001 law and only one—Albus—under the 1997 law.

factors (market size and proximity to the source country) are the most important determinants of FDI. The gravity model consistently explains about 60 percent of aggregate FDI flows, regardless of the region. Since gravity factors are exogenous, this finding puts into perspective the efforts of policymakers in host countries to attract FDI." One policy implication is that attracting FDI into these industries can turn into a race against neighboring countries, as country-level investment strategies of global companies are contingent on past decisions to establish production facilities serving the same markets. Of course, host-country policies can counteract such path-dependent advantages, especially those policies that affect relative unit labor costs, the corporate tax burden, infrastructure, and the trade regime.

From a review of the literature, we can identify several reasons why M&A FDI could lead to follow-on Greenfield investment: first, a Brownfield investment gives companies a foothold on which to grow by adding manufacturing capacity for the same or new product lines; second, the Brownfield acquisition is a positive signal to other foreign investors regarding the economic prospects for entry in a given product market and region; and third, M&A FDI can raise the confidence in the investment climate as a whole. There is a countervailing argument mentioned in the industrial organization literature—that M&A FDI could deter entry if the first investor obtains an essential facility, substantial excess capacity, or other assets it can use for predatory competition (i.e., allowing the company to set prices temporarily below variable costs).

Calderon, Loayza, and Serven (2004) study FDI flows to developing countries, which surged in the 1990s to become a leading source of external financing. This rise in FDI volume was accompanied by a marked change in its composition: Investment taking the form of acquisition of existing assets (M&A) grew much more rapidly than investment in mainly new assets (Greenfield FDI), particularly in countries undertaking extensive privatization of public enterprises. This raises the question of whether the M&A boom is a one-time effect of privatization, or is likely to be followed by a rise in Greenfield investment. The study finds that in developing and industrial countries, higher M&A is typically followed by higher Greenfield investment; the reverse holds true for industrialized economies, where Greenfield investment leads to M&A. In addition, domestic investment leads to both types of FDI in developing economies; in the case of industrial countries, domestic investment leads to M&A FDI, but is preceded by Greenfield FDI.

When considering the issue of sequencing of acquisition and Greenfield FDI, it is worth bearing in mind that the distinction is not clear-cut. Our case study illustrates the point that in transition economies, Brownfield FDI has been used as a vehicle for investments that should actually be considered as Greenfield investments. Buyers discard all or most production equipment, gut and recondition the plants, introduce imported machinery and equipment for new product lines, and, simultaneously, undertake extensive layoffs and new hiring. Consequently, by the time the restructuring is complete, little of the original plant or workforce remains operational.[42] This gives rise to the question of why an

---

42. Another example in which this distinction blurs is reported in the literature, and concerns the opposite case, where a Greenfield is simply a legal entity that takes over the assets of an existing company, as part of a bankruptcy, for tax purposes, or to avoid certain liabilities.

investor would decide to bear the extra costs associated with a difficult restructuring when a Greenfield investment could permit more flexibility in employment, plant layouts, and so forth.

Among the answers to this question, the presence of artificial entry barriers figures prominently. In all manufacturing industries included in the study, our interviews with managers indicated that there are important constraints confronted by Greenfield FDI that involve delays in obtaining licenses, such as building permits, from local and national authorities. The *Doing Business 2007* indicators confirm that Serbia continues to be far behind in this area: as an example, the report estimates that it takes 20 licensing and other procedures to build a warehouse, requiring 211 days and the associated costs are close to twenty times the country's average income per capita.[43]

## Post-acquisition Incentives to Increase Capabilities: The Role of Corporate Governance and the Investment Climate

Absorption of technology and the introduction of organizational innovations require risk taking, the willingness to accept a redistribution of tasks, responsibilities and relative compensation among employees, and the financial capacity to compensate workers facing technical redundancy. In the literature on insider control, we find many reasons why such behavior would not develop in the absence of a new private owner with a controlling interest, whether domestic or foreign. In Russia, Desai and Goldberg (2001) discuss how insider-dominated management blunts the incentives to improve competitiveness, and instead leads to asset stripping and accumulation of wage, supplier, and tax arrears. Political economy models similarly highlight the importance of corporate governance failures for absorption, as vested interest groups whose rent would be eroded by the adoption of superior technologies have strong incentives to block productive investments (Parente and Prescott 1999, 2000; Bridgman, Livshits, and MacGee 2007, and references therein).

Is there strong firm-level demand for absorption and innovation in ECA? Probably the most extensive case studies of FDI in the CEE are those in the books edited by Estrin, Richet, and Brada (2000), and by Moran, Graham, and Blomstrom (2005). Estrin, Richet, and Brada (2000) conducted case studies of four Slovenian, four Czech, and four Bulgarian enterprises in the late 1990s. The most important conclusion is that the pace of restructuring depends on the power of insiders, the nature of the product, and the market structure. They find that cost factors are important in the case of intermediate products, while strategic and market factors are crucial in the production of final goods. Unsurprisingly, strategic factors are important when the foreign firms operate in oligopolistic markets. This finding supports the theoretical arguments of Aghion, Dewatripont, and Rey (1999), who show that "competition, combined with the threat of liquidation, acts as a disciplinary device [for managers], which fosters technology adoption and growth." Although conservative managers would prefer to delay technology adoption, competition forces

---

43. See http://www.doingbusiness.org/ExploreTopics/DealingLicenses/Details.aspx?economyid=206 for details about these calculations for the case of Serbia.

them to upgrade just to break even, and this decision, in turn, increases competitive pressure for rivals. The findings point to some important consequences of FDI as we observe that acquisitions were followed rapidly by an upgrading of product quality and manufacturing methods.

A recent study has observed that poor performance and lack of innovative investments by Macedonian firms might be linked to low ownership concentration, which emerged from the mass (voucher) privatization process in the Former Yugoslav Republic of Macedonia. Where institutions are weak, concentrated ownership is required for investors to have sufficient incentives to enter and incur the sunk costs of absorption because only a high concentration of ownership can protect the investor from corruption, hold-up strategies by vested interests, and an unfriendly investment climate.[44] Thus, to compensate for weak institutions, strategic investors demand a majority interest, or, at a minimum, operating control (Goldberg and Nellis 2007; Goldberg and Radulovic 2005). In very weak institutional regimes, investors might require protection against a blocking minority or pay for guarantee instruments to guard their operation from unmanageable risks.

The case studies in this chapter also furnish evidence that demand for new products and changes in product mix are motivated by the type of ownership and a resulting change in corporate governance. In the case of Serbia, ownership and corporate governance are primarily the result of two laws: the 1997 law that gave away shares to managers and employees, and the 2001 Law on Privatization, which stipulates that 70 percent of the shares in each enterprise will be sold to a strategic investor using a competitive process, with the objective of reserving a legal controlling majority for a core investor. The capital that could be acquired free of charge by employees was stipulated to be no more than 15 percent of the capital of the enterprise. Later in this chapter we will discuss the effects of this policy on corporate governance and the resulting demand for innovation and technology absorption.

Managerial skills are an essential ingredient in this transformation of corporate governance. Foreign investors that acquire Brownfield manufacturing plants can rely on different solutions to address the lack of managerial capabilities. The main ones are: (i) the replacement of top management by foreign staff already employed by the MNE in the case of strategic investors; (ii) recruiting new management working for foreign *or* domestic companies operating in related *or* different industries; and (iii) developing younger staff to take more responsibilities. As we discuss further, all three solutions were present to varying degrees in each of the companies, and the combination selected seemed to depend as much on the corporate culture of the buyer as on the state of the management at the time of acquisition.

Appropriability concerns can be a barrier to developing local managerial and technical capabilities. Training a younger cohort of domestic managers and workforce with general skills and industry-specific skills, all of which are directly relevant for competitors, can pose a problem unless the incentive structures motivate and help retain personnel. Research shows that manager turnover is higher after acquisition, and such turnover can

---

44. For example, there is evidence of negative correlations between institutional development variables—risk of expropriation and the rule of law—and R&D expenditures from panel data analysis (Clarke 2001).

include talented managers with better external employment opportunities (Walsh and Elwood 2001).[45] Turnover rates tend to be higher when the buyer is foreign, and in part this reflects differences in organizational culture and structure (Krug and Hegarty 1997). As we describe below, one constraint that investors faced in Serbia was that workforce reductions had to be accomplished via general severance packages that were often accepted by highly capable workers, who were the ones who could more easily find employment elsewhere.

## The Background of the Serbian Privatization Program

In this section, we outline the context for the acquisition and investment decisions in the case study companies. The core of Serbia's enterprise sector reform strategy has been the ambitious program of privatizing socially-owned enterprises (a collective form of property ownership that is controlled by the employees). Serbia's 2001 privatization law (amended in 2003) incorporated international best practices and lessons learned from a decade of experience in other transition economies. The Privatization Law stipulates three methods of privatization: (i) *tenders* of large enterprises, offering to a strategic investor at least 70 percent of the shares; (ii) *auctions* of medium-sized enterprises; and (iii) *restructuring* and subsequent tenders and auctions of a select group of large, presently loss making, but potentially viable enterprises, or parts thereof.[46]

After five years, 1,407 enterprises have been privatized through competitive public tender and auction procedures, with privatization proceeds reaching nearly 1.7 billion euros, and social and investment program commitments of almost 1.4 billion euros. The fact that 74 percent of all offered companies were actually sold is impressive considering the numerous challenges resulting from the legacy of social ownership.[47] Note that the total sales value of companies sold through public tender is approximately the same as the *ex ante* investment committed in the bids. The compromise between immediate privatization revenues and the long-term sustainability of companies that was agreed upon by government and company stakeholders is a significant feature of the privatization process.[48]

There were also about 800 to 1,000 companies that were privatized according to the 1997 Privatization Law, prior to February 2001, when the post-Milosevic government took power. The 1997 law gave away 60 percent of a company's shares to employees free of charge, and 30 percent were offered for sale to the insiders at a deep discount and in installments. According to a 2005 report financed by the European Agency for Reconstruction titled *Impact Assessment*

---

45. This problem is economically relevant, given that the direct recruitment costs for replacing managers are estimated to be around 50 percent of salary (Development Dimensions International 2003), on top of which there are likely to be substantial adjustment costs.
46. The analysis of privatization in this chapter is based on Goldberg and Radulovic (2005).
47. Data source: Privatization Agency 2007.
48. In addition to the stake price, in the tender privatizations the acquiring companies committed to specified investments as well as social programs, so in many cases investors were obliged to invest substantial amounts in the post-privatization period.

Table 4.1. Results of the Serbian Privatization Program

| No. of Companies Sold | 2002 | 2003 | 2004 | 2005 | 2006 | '02-'06 |
|---|---|---|---|---|---|---|
| Tenders | 11 | 16 | 8 | 15 | 25 | 75 |
| Auctions | 165 | 560 | 214 | 185 | 208 | 1,332 |
| Total | 176 | 576 | 222 | 200 | 233 | 1,407 |
| Revenues (in 000s euro) | 2002 | 2003 | 2004 | 2005 | 2006 | '02-'06 |
| Tenders | 200,691 | 594,748 | 11,395 | 96,516 | 101,202 | 1,004,552 |
| Auctions | 42,214 | 205,070 | 109,157 | 164,100 | 160,888 | 681,429 |
| Total | 242,905 | 799,818 | 120,552 | 260,616 | 262,090 | 1,685,981 |
| Share Fund (SF) | 82,968 | 67,754 | 51,938 | 124,828 | 67,727 | 395,215 |

*Source:* Privatization Agency, in Cvetkovic, Pankov and Popovic, "Balkan Late Comer—The Case of Serbian Privatization," in Lieberman and Kopf (2007).

*of Privatization in Serbia,*[49] companies privatized under the 2001 law have generally improved their financial performance and have invested in modernizing their production process, while companies privatized to employees based on the 1997 law showed on average poorer financial results and lacked significant investments in modernization. Additionally, the cases where the employee shares were later bought by major shareholders (investment funds or strategic investors) showed much better performance than those with employee ownership, which adds evidence concerning the inferiority of employee-ownership privatization.

## Comparative Case Histories of Post-privatization Absorption

Eight large companies operating in the metal processing, household chemical, pharmaceutical, and cement industries were selected for the case study. In brief, the guiding selection criteria were that: industries, as well as company characteristics (especially the size of the firm), were useful for the *comparability* of results; company characteristics and the type of acquisition (especially the type of buyer), provided some *controls to test counterfactuals;* companies were privatized early on to ensure the availability of archival information (due diligence, post-acquisition monitoring), and sufficient time for key restructuring and investment decisions to have been implemented. For a longer description of the methodology used to carry out the case study, see Box 4.1. In this section we proceed by presenting the main results regarding absorption via acquisition FDI in different industries, first by setting out the highlights of analysis using financial information for all the companies, and then comparing the pairs of companies in the metal processing and household chemical industries across key dimensions of the post-acquisition restructuring.[50]

---

49. "Impact Assessment of Privatization in Serbia"—report prepared by the IDOM/SEECAP consultant team (Annie Cordet-Dupouy and Jaime Temes) as a part of the "Preparation of Companies for Privatization" project, EuropeAid/116898/D/SV/YU.

50. A forthcoming working paper dealing with the case study will present the full set of results from the financial analysis and interviews, which for reasons of space it is not possible to include in this report.

### Box 4.1: Methodology: Company Selection, Data Sources, Fieldwork

The industries and company characteristics were selected to allow *comparability* of results. Eight companies were selected for the case studies belonging to the following industries: metal processing, household chemical, pharmaceutical, and cement. The focus on larger companies (at privatization, all companies had at least 500 employees) operating in traditional manufacturing industries tries to ensure that the investment decision matrix faced by potential investors is as similar as possible. Due diligence documents we use to reconstruct the "baseline" conditions of the companies certainly indicate that all companies were running at low capacity, had been undercapitalized for several years, suffered from having machinery that was old and even out of service, etc.

A selection criterion to maximize comparability and the availability of pre-acquisition data is that companies with FDI were all privatized through the tender process. The tender process (see details in next subsection) provided an impartial framework within which investors were able to obtain full and accurate information about the companies, and offer bids that included a detailed and transparent strategy for investment and treatment of the workforce. The acquirers were monitored to ensure compliance regarding purchase conditions. All foreign investor-owned companies we include were sold by this method in 2002–03. Therefore, the rules and horizon for investment are very similar. Thanks to the access that we had to due diligence material, we were able to check the pre-acquisition information about all characteristics of the companies.

The selection criteria introduced proper *controls to test counterfactuals*. Since an objective of the study is to pinpoint how the acquisition with FDI differs from that involving domestic investment, we chose pairs of companies with different types of owners in as far as this was possible. The two metal processing companies were bought by a regional MNE from Slovenia and a domestic investor. One household chemical company was bought by a large German-based MNE, another was bought by a domestic investor, and a third is yet to be privatized. The pharmaceutical companies were bought by a large Icelandic MNE and by insider management. Finally, one cement company bought by a large French MNE was included. No controls were available in this industry, as all three major cement producers received FDI from large MNEs based in different EU countries.

The archival information includes all pre-acquisition documents prepared during privatization, and post-acquisition monitoring reports of the Serbian Privatization Agency. The access to archival information from the Privatization Agency was invaluable in selecting the companies and understanding the baseline from which investors had to build up capacity and capabilities in many dimensions. Moreover, the pre-acquisition documents give details of the *ex ante* strategic investment plans agreed upon by the investors in their bids, and we are therefore in a position to detect divergences on the upside or downside for each item (whether repairs, new equipment, ICT, training, etc.), by comparing these intentions with the monitoring reports.

Annual financial statements and in-depth interviews with top management were the two fundamental sources of information about the companies' evolution following the acquisition. A visit to Serbia to carry out fieldwork took place in March 2007. During this visit, a team of Bank specialists, including the author, visited the companies to meet the CEO and directors who were responsible for investment, technical control, finance, sales, and marketing. The two- to three-hour-long discussions were based around an open-ended questionnaire prepared in consultation with experts of the privatization process in Serbia; the questionnaire was sent in advance to the companies in English and Serbian versions (the questionnaire we used is presented in Appendix C). The interviews with managers covered several areas including:

- *Innovation and investment:* such as, whether new and improved products and manufacturing processes introduced were developed by the enterprise, developed together with other enterprises or institutions, transferred by the enterprise group, or acquired externally; the costs and time taken by the development or transfer, and how this process was organized (appraisal, decision, implementation, evaluation); the arrangements to transfer technology (origin, cooperation partners); and the investment plan signed at the time of acquisition.
- *Standards and quality certification*
- *Skills and training*

*(continued)*

> **Box 4.1: Methodology: Company Selection, Data Sources, Fieldwork** *(continued)*
>
> - *Business practices and management:* such as, IT-based management information systems, cross-functional teams, outsourcing of business functions, disposal of non core activities, rationalization of distribution channels, changes in marketing strategies, reporting lines and frequency, changes in executive directors and board members.
> - *Competition and exports*
> - *Sources of government support*
>
> Meetings with current and former government officials and industry experts were an important complementary source of information. As part of the visit to Serbia, the team met with current and former officials of the Privatization Agency and of the Share Fund of the Republic of Serbia. These meetings offered numerous insights into institutional issues surrounding the privatization process, and specifically regarding the monitoring of privatized companies and ongoing efforts to sell remaining socially owned enterprises. Discussions with faculty members at Belgrade University contributed valuable information regarding training and skills in the industries in question.

## *Overview of Companies and Trends in Post-acquisition Results*

Several common trends exist when we compare pre-acquisition and post-acquisition levels of operating income, employment, and salaries. As shown in Table 4.2, acquisition is synonymous with a robust increase in operating income and a concurrent reduction in the workforce, both of which show up clearly within the initial two- to three-year time period, with FDI acquisitions showing more pronounced changes overall. The combined effect is a substantial jump in operating income per worker, ranging from 24 percent for the household chemical company bought by a local investor, to 308 percent for the aluminium company bought by a foreign investor, with companies receiving FDI showing significantly larger gains. Salaries for those workers that remain following retrenchment and restructuring rise in line with efficiency improvements, and are between 99 percent and 170 percent higher for FDI acquisition companies. In stark contrast, the company that is yet to be privatized has seen a fall in operating income relative to the year when it should have been privatized, and salaries have stagnated.

Trends in firm-level productivity, shown in Table 4.3, are generally consistent with the hypothesis that acquisition by a foreign investor brings about more efficiency-enhancing factor reallocation, restructuring, and investment, compared to local investors. Generally speaking, the leading driving force behind this change in labor productivity is downsizing of the workforce, as hiring and firing rigidities are removed by the introduction of new ownership arrangements. This is consistent with the view that privatization to strategic buyers permits labor shedding in enterprises that were subject to substantial labor hoarding. Our interviews with CEOs suggest that value added is likely to rise further as the companies turn from rehabilitating to purchasing machinery and equipment, the lead time for installing new capacity is completed, and the effects of learning by doing show up in the financial data. In the same period, the productivity of the non privatized household chemical company fell significantly, as overall value added declined.

Although the financial results show changes in operating income, salary, and productivity of the same sign and approximately of the same magnitude across the companies

Table 4.2. Revenue and Employment Trends Pre- and Post-acquisition[51]
(Income and salary stated in thousand current US$)

| Company | Period | Number of Employees | Operating Income | Total Operating Income Per Employee | Salary Per Employee | Acquisition Details |
|---|---|---|---|---|---|---|
| Impol-Seval | Before | 1,111 | 27,239 | 25 | 3.32 | Privatized Oct. 2002 |
| Strategic regional buyer | After | 752 | 67,368 | 101 | 8.95 | Buyer Impol Slovenia |
| Aluminium processing | Change | −32% | 147% | 308% | 170% | Stake price 3.5 mln € |
| Nissal | Before | 1,179 | 11,536 | 10 | 3.38 | Privatized Sep. 2003 |
| Strategic local buyer | After | 914 | 19,863 | 25 | 5.54 | Buyer Domal Inzenjering |
| Aluminium processing | Change | −23% | 72% | 152% | 64% | Stake price 0.3 mln € |
| Merima-Henkel | Before | 1,122 | 34,527 | 31 | 4.93 | Privatized Oct. 2002 |
| Strategic foreign buyer | After | 815 | 62,670 | 89 | 11.36 | Buyer Henkel Germany |
| Household chemicals | Change | −27% | 82% | 188% | 130% | Stake price 14.4 mln € |
| Albus | Before | 539 | 8,626 | 16 | 3.81 | Privatized 2005 |
| Strategic local buyer | After | 423 | 8,097 | 20 | 4.92 | Buyer Invej Holding |
| Household chemicals | Change | −22% | −6% | 24% | 29% | Stake price 3.5 mln € |
| Nevena | Before | 745 | 7,171 | 10 | 2.87 | Not privatized yet |
| Unsold company | After | 633 | 5,098 | 8 | 3.18 | Tender failed in 2003 |
| Household chemicals | Change | −15% | −29% | −18% | 11% | |
| Zdravlje-Actavis | Before | 1,954 | 36,359 | 19 | 6.20 | Privatized Feb. 2003 |
| Strategic foreign buyer | After | 1,348 | 43,351 | 33 | 12.37 | Buyer Actavis, Iceland |

*(continued)*

---

51. In this and subsequent tables, the pre-acquisition averages are calculated from the financial statements from 2001 up to the year of privatization, which is different for each company; post-acquisition figures are inclusive of the privatization year and up to 2006. One of the main challenges of using these financial data was the change in standard accounting practices in these companies following privatization.

Table 4.2. Revenue and Employment Trends Pre- and Post-acquisition[51]
(Income and salary stated in thousand current US$) (continued)

| Company | Period | Number of Employees | Operating Income | Total Operating Income Per Employee | Salary Per Employee | Acquisition Details |
|---|---|---|---|---|---|---|
| Pharmaceuticals | Change | −31% | 19% | 77% | 99% | Stake price 3.5 mln € |
| Zorka-Hemofarm | Before | 645 | 26,487 | 41 | 6.14 | Privatized Dec. 2002 |
| Strategic local buyer | After | 577 | 32,337 | 56 | 13.51 | Buyer Hemofarm |
| Pharmaceuticals | Change | −11% | 22% | 37% | 120% | Stake price 14.6 mln € |
| Beocin-Lafarge | Before | 2,058 | 51,183 | 25 | 4.34 | Privatized March 2002 |
| Strategic foreign buyer | After | 1,126 | 74,568 | 76 | 10.47 | Buyer Lafarge, France |
| Cement industry | Change | −45% | 46% | 199% | 141% | Stake price 50.8 mln € |

Source: Authors' calculations based on the annual Financial Statements of the companies 2001–06, Solvency Center, National Bank of Serbia.

receiving FDI, the details of how this story played out was different in each, and responded to the conditions on the ground in manufacturing plants, the domestic market, and export potential for individual product lines, industry-specific best practices, etc. This heterogeneity in terms of the restructuring and operating strategy is evident from the wide range of values for standard financial ratios in Table 4.4. Take, for example, the current ratio, which is an indicator of whether short-term assets can cover immediate liabilities. This shows no common trend, let alone a convergence toward a given value, and this is true also of the management rate of return (operating income to operating assets), inventory turnover (cost of goods sold to inventory). One indicator that shows more consistent trends is debt to equity, which increases more for companies acquired by local investors, illustrating the point that they are more likely to obtain financing with domestic banks, rather than rely on equity injections and debt by acquiring MNEs in international capital markets.

## Aluminum Processing Industry

In this industry we compare in detail the post-acquisition evolution of a company bought by a regional strategic investor to one bought by a local non strategic investor.

### Table 4.3. Productivity Trends Pre- and Post-acquisition[52]
(Value added in current US$)

| Company | Period | Value Added | Number of Employees | Value Added Per Employee |
|---|---|---|---|---|
| Impol-Seval | Before | 2,818.91 | 1,111 | 2.55 |
| Strategic regional buyer | After | 3,696.06 | 752 | 5.52 |
| Aluminium processing | Change | 31% | −32% | 117% |
| Nissal | Before | 4,543.26 | 1,179 | 3.86 |
| Strategic local buyer | After | 3,855.74 | 914 | 4.30 |
| Aluminium processing | Change | −15% | −23% | 12% |
| Merima-Henkel | Before | 8,352.53 | 1,122 | 7.45 |
| Strategic foreign buyer | After | 12,781.81 | 815 | 17.38 |
| Household chemicals | Change | 53% | −27% | 133% |
| Albus | Before | 2,359.42 | 539 | 4.41 |
| Strategic local buyer | After | 2,361.68 | 423 | 5.77 |
| Household chemicals | Change | 0% | −22% | 31% |
| Nevena | Before | 2,961.83 | 745 | 3.98 |
| Unsold company | After | 2,021.24 | 633 | 3.09 |
| Household chemicals | Change | −32% | −15% | −22% |
| Zdravlje-Actavis | Before | 24,087.64 | 1,954 | 12.56 |
| Strategic foreign buyer | After | 26,632.29 | 1,348 | 20.23 |
| Pharmaceuticals | Change | 11% | −31% | 61% |
| Zorka-Hemofarm | Before | 12,891.99 | 645 | 19.99 |
| Strategic local buyer | After | 14,050.12 | 577 | 24.60 |
| Pharmaceuticals | Change | 9% | −11% | 23% |
| Beocin-Lafarge | Before | 14,241.96 | 2,058 | 7.19 |
| Strategic foreign buyer | After | 28,526.41 | 1,126 | 29.62 |
| Cement industry | Change | 100% | −45% | 312% |

*Source:* Authors' calculations based on the annual Financial Statements of the companies 2001-06, Solvency Center, National Bank of Serbia.

Valjaonica Aluminjuma, Sevojno (Seval), with sales of US$19 million and 1,140 employees in 2001, manufactures aluminum rolls, profiles, and semifinished and finished products for the construction industry. The company was sold by tender privatization in October 2002 to Impol d.d. from Slovenia, a large producer of aluminum products. Impol paid US$6.5 million for 70 percent of the social capital of the enterprise, and made US$14.6 million in investment commitments. The Information Memorandum at the time

---

52. Value added is defined as sales minus the cost of goods sold, raw materials, other material, and operating expenses.

## Box 4.2: A Closer Look at Productivity Trends

Underlying the value-added productivity increases at the firm level are impressive increases in physical productivity in core product lines. Owing to the confidentiality of the data, we do not identify the names of the companies or product lines in the tables below, which compare the annual physical productivity data for an FDI-receiving company and one bought by a local non-strategic investor in the same industry.

There are several interesting results. First, the overall physical productivity diverges significantly in the period following the acquisition, from being roughly similar, to being around four times greater in the company bought by the foreign investor. This is consistent with having more productivity-related investment and absorption when ownership goes to a strategic investor. Second, the divergence is even greater when one examines individual product lines. Here, the difference grows to close to eight times larger in the company that received FDI, supporting the hypothesis that foreign investors have incentives to focus on core product lines and close others, even if adjustment costs are greater.

Physical productivity in FDI acquisition company (physical units per company employee)

| Product Line | 2001 | 2002 | 2003 | 2004 | 2005 | 2006 | Before | After | Change |
|---|---|---|---|---|---|---|---|---|---|
| #1 | 4.1 | 7.3 | 14.8 | 28.8 | 49.4 | 39.7 | 5.7 | 28.0 | 389% |
| #2 | 0.1 | 0.7 | 0.2 | 1.1 | 1.2 | 1.4 | 0.4 | 0.9 | 128% |
| #3 | — | — | — | 6.3 | 9.9 | 14.2 | — | 10.1 | — |
| #4 | 3.0 | 0.4 | 13.1 | 11.1 | 13.1 | 27.0 | 1.7 | 13.0 | 658% |
| #5 | 0.7 | 0.6 | 0.6 | 0.6 | 0.6 | 0.3 | 0.6 | 0.5 | −17% |
| #6 | — | — | — | 0.4 | 1.9 | 2.7 | — | 1.6 | — |
| Total | 8.0 | 12.7 | 28.8 | 48.2 | 76.1 | 85.3 | 10.3 | 50.2 | 386% |

*Source:* Authors' calculations based on information provided by the company.

Physical productivity in company bought by local investor (physical units per company employee)

| Product Line | 2002 | 2003 | 2004 | 2005 | Before | After | Change |
|---|---|---|---|---|---|---|---|
| #1 | 2.9 | 3.0 | 5.5 | 5.7 | 2.9 | 4.7 | 61% |
| #2 | 2.2 | 2.5 | 5.0 | 5.9 | 2.3 | 4.5 | 92% |
| #3 | 0.8 | 1.3 | 2.3 | 2.4 | 1.1 | 2.0 | 89% |
| #4 | 0.4 | 0.5 | 1.6 | 2.7 | 0.5 | 1.6 | 236% |
| #5 | 0.2 | 0.1 | 0.3 | 0.3 | 0.1 | 0.2 | 60% |
| #6 | 0.2 | 0.3 | 0.4 | 0.5 | 0.2 | 0.4 | 57% |
| #7 | — | — | 0.4 | 0.5 | — | 0.3 | — |
| Total | 6.7 | 7.7 | 15.4 | 18.1 | 7.2 | 13.7 | 91% |

*Source:* Authors' calculations based on information provided by the company.

Table 4.4. Financial Ratios Pre- and Post-acquisition

| Company | Period | Working Capital Ratio | Management Rate of Return | Current Ratio | Inventory Turnover | Debt to Equity |
|---|---|---|---|---|---|---|
| Impol-Seval | Before | 1.14 | 2.94 | 1.14 | 3.91 | 0.95 |
| Strategic regional buyer | After | 0.81 | 3.18 | 0.81 | 5.75 | 2.13 |
| Aluminium processing | Change | −29% | 8% | −29% | 47% | 123% |
| Nissal | Before | 2.51 | 1.13 | 2.51 | 3.18 | 0.24 |
| Strategic local buyer | After | 1.65 | 1.44 | 1.65 | 3.46 | 0.66 |
| Aluminium processing | Change | −34% | 27% | −34% | 9% | 178% |
| Merima-Henkel | Before | 4.08 | 1.59 | 4.08 | 6.77 | 0.13 |
| Strategic foreign buyer | After | 4.20 | 1.52 | 4.20 | 11.69 | 0.14 |
| Household chemicals | Change | 3% | −4% | 3% | 73% | 5% |
| Albus | Before | 1.49 | 0.77 | 1.49 | 4.26 | 0.57 |
| Strategic local buyer | After | 1.01 | 0.60 | 1.01 | 2.60 | 0.69 |
| Household chemicals | Change | −32% | −22% | −32% | −39% | 20% |
| Nevena | Before | 1.23 | 0.97 | 1.23 | 3.35 | 0.20 |
| Unsold company | After | 0.73 | 0.97 | 0.73 | 3.68 | 0.78 |
| Household chemicals | Change | −41% | 0% | −41% | 10% | 290% |
| Zdravlje-Actavis | Before | 2.57 | 0.97 | 2.57 | 4.92 | 0.21 |
| Strategic foreign buyer | After | 3.99 | 0.72 | 3.99 | 4.97 | 0.48 |
| Pharmaceuticals | Change | 55% | −26% | 55% | 1% | 132% |
| Zorka-Hemofarm | Before | 7.25 | 1.04 | 7.25 | 2.87 | 0.06 |
| Strategic local buyer | After | 4.98 | 0.68 | 4.98 | 3.18 | 0.14 |
| Pharmaceuticals | Change | −31% | −34% | −31% | 11% | 110% |
| Beocin-Lafarge | Before | 0.71 | 1.77 | 0.71 | 6.43 | 0.54 |
| Strategic foreign buyer | After | 0.76 | 1.25 | 0.76 | 4.51 | 2.03 |
| Cement industry | Change | 8% | −29% | 8% | −30% | 275% |

*Source:* Authors' calculations based on the annual financial statements of the companies 2001–6, Solvency Center, National Bank of Serbia.

of acquisition states that "Seval is a *well-run company,* which has successfully survived the political and economic disruption . . ." (page 8, italics added), and that Seval's plant was kept in a "neat and tidy condition and shows clear evidence of the stated policy of maintaining a high quality standard." Despite these claims, Impol found that there had been virtually no investment in the production facilities for 10 years, there were problems with internal transport of material and goods, and that, "computer systems are well-behind the position of systems in a typical Western company."

Nissal, a company located in the city of Nis, had sales of US$10.5 million and 1,250 employees in 2001. It produces aluminium profiles, rod, wire, and semifinished and finished products for the construction industry. The company was sold by tender privatization (in a second attempt) in September 2003 to Belgrade-based Domal Inzenjering for €0.3 million, for 70 percent of the social capital stake and a €3.1 million investment commitment. An Information Memo by the same financial adviser (Fieldstone) states that the company successfully survived the political and economic disruption but, in contrast to Seval, does not characterize Nissal as being a well-run company.

On the surface, the difference in the price paid for the companies is surprising: Seval, with sales only twice as high, commanded a 20 times higher price. Can the explanation lie in the difference in efficiency between the companies at the time of privatization, if the measured operating income per employee is only two and a half times higher at Seval than at Nissal? Interviews with the financial advisers that supported Seval's acquisition indicate that the differential was due to the efficiency and potential revenue of the company. Seval's equipment was very well-maintained, its workers were better trained and more highly qualified, and the company was outsourcing whatever was cheaper to buy, instead of producing it, whereas Nissal had a highly integrated production process, which translated into stronger unions and workers trained to produce the "wrong" products from the investors' viewpoint.

In addition, the financial advisers drew attention to market-related and intangible factors as a second set of reasons for this price premium. First, whereas the management at Seval was very cooperative and supportive in the privatization process, since they wanted the acquisition to go through and they knew the investor from prior business dealings, the Nissal management obstructed the privatization process from the very beginning, which led to a failure of the first tender. Second, Seval's product lines were designed for export and suitable for a wide market, while at Nissal, the products were mainly designed for the domestic market, including some for military purposes, and therefore these products were not suitable for the typical EC consumer market. Third, there was strong interest from Impol, because it had to either close down its production facility in Slovenia, or renovate it (an option that would have been more expensive). After the failure of the first tender for Nissal due to a lack of cooperation by the Company in the process of privatization and a lack of investor interest, a second tender was characterized as lacking competition, but there was great pressure to get an investor on board.

*Restructuring and Investment.* Both companies took steps to restructure by rationalizing their product mixes. Following the acquisition of Seval by the Slovenian investor, an exchange of equipment between plants in Serbia and Slovenia took place, in order to switch product lines (through sale-purchase between the companies); for example, foil is now produced only in the Slovenia plant. In the case of Nissal, the new owner closed the production lines for irrigation tubes and solar collectors, manufactured manually in small batches and generating no profit. Since privatization, Nissal has begun producing a new product (windows), transferred by the buyer, who has been producing windows at his other business. Since 2003, a new product—profiles—is produced in Nissal under license from an Italian company.

In both companies, restructuring translated into the closure of product lines that were not breaking even, or were produced in small batches. The renewed search for profits since privatization had different consequences under foreign or local management, although we cannot generalize from two examples. In the case of FDI, the emerging pattern is regional specialization of production by switching product lines across the investor's active manufacturing sites, responding to economies of scale. For the domestic investor, where this is not possible, the search led to the introduction of higher-value product designs, where there are economies of scope. In the case of Nissal, this extension involves new manufacturing under license to foreign firms, which transfer the know-how and give branding advantages.

Since the privatization of Seval, over €31 million has been invested. The investments made by Impol-Seval have largely consisted of purchases of new machinery and equipment imported from more advanced economies to increase productivity and expand capacity, reducing environmental impacts. Out of the €16 million invested in 2006, €11 million was used to build a new casting complex to cast rolling ingot, doubling the production level to 100,000 tons. The casting equipment came from the United States, and other big investments involved imports of machinery and equipment from Germany, Belgium, Austria, and Finland. Only the IT was partially domestic in origin. Consequently, it is fair to say that FDI acquisition generated technology absorption in Seval via a process of technology transfer that has relied on importing capital goods embodying the knowledge at the global technology frontier. Since the privatization of Nissal in 2003, €3 million has been invested, mainly in refurbishment and replacement of outdated machinery. Plans are currently underway for a €5 million new production line to manufacture aluminum alloy pipes for export, with a capacity of 18,000 tons.

*R&D and Quality Certification.* Since acquisition Seval has cut back on its technical and quality control, R&D and related costs by around 75 percent since acquisition, although in 2004–06 it still reported working on 88 projects, and 19 more were being undertaken in 2007. Projects mostly aim to resolve *production-related problems*. Management reports that there is a more structured R&D effort, aimed at improving the production process for new products, such as the development of hot-rolled plates using alloys with magnesium for use in the shipbuilding industry. Two full-time people work on R&D projects, and there is cooperation with universities in Belgrade and Ljubljana. Concrete examples of this cooperation include joint research into aluminum alloys, leading to co-publications in local and international scientific journals (Romhanji and others 2006), and sponsorship of specialized metallurgy conferences in which Seval staff also participate.

In Nissal, 20 employees worked in the R&D department prior to privatization, using old and outdated equipment in the laboratory. Post-privatization, the department was downsized and new, specialized employees were hired, but we were not able to learn about their capabilities or the specifics of the R&D program.

Seval was accredited to ISO 9002/1994 in 1999, well before its 2002 privatization. In 2003, Seval was awarded an ISO 9001:2000 certificate, renewed in January of 2007, and which is valid through October 2009. While Nissal had no certification prior to its 2003 privatization, it is now ISO 9001 certified.

*Labor, Skills, and Training.* The workforce has been reduced from 1,140 in the year of acquisition to 740 most recently at Seval, and from 1,250 to 650 at Nissal (with an additional 150 employees being scheduled for retirement within 2007–08). A new incentive structure was put in place at Seval after privatization, which makes it possible for two engineers in the same position to have a 100 percent difference in salary, depending on performance. In Nissal, a performance-based reward system applies to employees and managers; for example, it was reported that a 10 percent to 60 percent bonus on base salaries could be received by employees.

Since its acquisition by Impol, Seval's workers were sent for training to enterprises located in Slovenia, United States, Bosnia and Herzegovina, and Italy. Seval also has an agreement on knowledge transfer with Impol, which allows it to send workers to the Slovenian facility for determined periods. The share of personnel with associate and university degrees has increased marginally, from 11 percent to 15 percent. Problems were reported in finding qualified metallurgical engineers, as there is little interest in this type of career. In contrast, Nissal reported minimal training: computer skills training for administrative staff and for two managers responsible for IT technology. This suggests that FDI sometimes brings with it an international knowledge "supply chain" that employees in acquired subsidiaries can tap into, whereas those receiving local capital remain constrained. We explore this further in other companies.

*Corporate Governance.* Before privatization, under social ownership in ex-Yugoslavia, Seval's executive management board and the general manager of the company were elected by the company general assembly, which included only the workers. After privatization, the board consists of local directors and representatives of Impol.

After privatization, management was given the freedom to make decisions in a more independent and commercially oriented way. The organizational structure changed significantly, largely influenced by Impol, with more decentralized decision making. Yet, even if local management is now in control of the decision-making process, production processes are under video surveillance, allowing managers in Slovenia to monitor developments on the production floor in real time. Transfer of business practices from Impol has been informal and achieved mainly through mutual interaction.

Before privatization, Nissal was managed by a general director who worked at the company for 22 years, but who exercised limited control over the affairs of the company owing to the corporate governance in socially owned enterprises. After privatization, the new owner changed the financial director, then the directors of five production units (plants), and finally appointed a new general director. Half of the new management team is from outside the company. The new general director is an engineer who worked under the old regime in the Ministry of Interior, and the deputy general director is an economist with experience in commercial banking.

*Investment Climate.* At the time of acquisition, Seval had 90 percent of the domestic market for its product mix. Since then, there has been a progressive closure of smaller product lines aimed at the domestic market (corrugated sheets), and the introduction of new product lines for export markets (hot-rolled plates and coils). It continues to have a dominant position domestically (over 50 percent for its main product line), with little competition from other firms. Greek and Croatian producers have entered the market, but they still have small market shares. It is the only producer of hot-rolled plates, while the other products compete with small foundries, importers, and substitute materials. Since the Slovenian investment, Seval secured access to new foreign markets through Impol, reestablishing the network of clients that Seval had lost in the 1990s.

Nissal's main products, aluminum profiles and bars, are exported, with 65 percent of its products going to Germany. Only 20 percent to 30 percent of the production goes to the local market. Nissal is the only company on the local market with a license from the Italian company AL Progetti & Consulenze for aluminum facades.

## Household Chemical Industry

In this industry we compare a company sold to a multinational strategic investor with another bought by a local non strategic investor.

Merima, based in the town of Krusevac, had sales of €30 million before privatization (2001), and it produced household cleaners, cosmetics, and personal hygiene products. The company was sold by tender privatization in the first attempt in October 2002 to Henkel, which paid 14.4 million for 70 percent of capital and made €43 million in investment commitments. The buyer is a company headquartered in Düsseldorf, Germany, that has about 52,000 employees worldwide; its three globally operating business sectors are laundry and home care, cosmetics and toiletries, and adhesive technologies.

Albus, founded in 1871, was privatized under the 1997 law, whereby 60 percent of the shares were given away to managers and employees, with the remainder remaining in state hands. The company was acquired through the capital market in February 2005, with the acquisition of 50.5 percent of the shares by the local holding company Invej. The holding company also has companies selling refrigerators and household goods, foodstuffs, and tobacco, among other consumer products. It is therefore not a strategic investor.

*Restructuring and Investment.* After Henkel acquired Merima, it immediately introduced changes in technology and formulas of all powder detergents. This is the main product line, and productive efficiency is now sufficiently high that Henkel has closed its detergent plants in Slovenia and Ukraine, and produces from Serbia instead. A line for production of fatty acids was shut down, and the plant was rebuilt to make way for craftsmen adhesives production. This product was new to the local market and new to Merima. Knowledge was transferred with the help of staff from Austria and Romania. The fatty acids production plant was renovated and, with the purchase of new machines, was transformed into a craftsmen adhesives plant with a capacity of 55,000 tons. Another plant for craftsmen adhesives is being built in Indjija (expected capacity 100,000 tons, €5.2 million investment, 60 workers).

Production of a cosmetics and toothpaste filling line was moved from Krusevac to Slovenia, including the machinery and equipment. New products were introduced in Serbia: compact detergents and window cleaners (Meriglass) based on nanotechnology. A premium detergent (Persil), which Merima produced under licensing agreements, is now being produced and marketed by Merima itself. This is another example where there is product line switching among other plants of the MNE and the host company, in order to take advantage of regional opportunities for manufacturing specialization.

At Albus, production stopped for six months after acquisition, and after reopening, all product lines were operative within a year. A toothpaste product line was discontinued. There were limited product introductions, consisting mainly of extensions of product lines (new dishwasher detergent) and improved presentations (e.g., packaging redesign). The investment targets increasing brand protection and awareness by registering old and new products in Serbia, and in more than a dozen neighboring countries to which it will be exporting. As for manufacturing technology, the company bought new packaging and labelling equipment in 2006, financed with a loan from the holding company. There are plans for purchasing a production line for softeners, which would be used later for liquid detergents.

*R&D and Quality.* Since privatization, the laboratory in the Krusevac plant stopped its more basic research because Henkel decided basic research was conducted more efficiently in Düsseldorf. The main activity nowadays is to adjust formulas from Henkel headquarters to allow production with existing local machinery. The lab performs testing and gives approval to raw materials from new suppliers, and tests competitors' products. Twelve people are employed in the R&D department. All innovative research is carried out by Henkel in Germany and in subsidiaries in France, Ireland, Japan, and the United States. Altogether, 2,800 employees work in research, product development, and applications engineering. Albus, on the other hand, continues to employ 23 technology staff in the development department, but it is unclear what their activities consist of, given limited product introductions (one new formula for detergent TWIST), or new machinery.

At the time of the privatization, Merima was not certified as ISO 9000. Since then, management reports that production has been organized so that it conforms to Henkel's safety, health and environmental standards, and therefore Henkel sees no need for additional ISO certification. Albus was issued an ISO 9001 certificate in 2000, which it finds important in order to export to the EU market. MNEs that form a tight relationship with their clients are required to meet company-specific standards rather than ISO standards. This is especially so in the areas of health and safety, where company image is at stake.

*Labor, Skills, and Training.* Currently, Henkel Merima employs 740 people, compared to 1,182 workers in 2000. A voluntarily layoff program was launched in 2003, and 400 employees left. The company further reduced the workforce by declaring 160 to 170 employees redundant. Since privatization, Henkel Merima has hired 35 to 40 young, skilled, and motivated workers. The company introduced performance-oriented incentive schemes. Henkel mainly organizes internal training, either in its regional headquarters in Vienna, Austria, or by sending functional managers to Krusevac to provide on-site training. The company has a job rotation practice. Managers receive corporate training. In order to be able to apply Henkel's own safety standards, employees had to be trained in occupational safety.

In the case of Albus, the number of employees was reduced from 523 to 410, with 120 employees opting for voluntarily leave and taking severance payments. Since 2005, a new recruitment process has been in operation for skilled, highly educated people in the procurement department and in sales. These employees have a good working knowledge of standard software. The company has introduced performance-oriented incentive schemes for employees, which can increase or decrease employees' salaries in case of good or bad performances, or in case of missed deadlines.

*Corporate Governance.* Before privatization, (according to the Information Memorandum prepared by Nomura for the tender), the highest decision-making body of Merima was an assembly consisting of 21 members, all employees of the company. Members of the assembly were elected by employees for a period of five years, with the possibility of reelection for a further five-year term. For 10 months after privatization, Henkel relied on the management team in Merima, resulting in delays in restructuring. A new general manager, a Serbian expatriate, was recruited from the competitor company Procter & Gamble, where he had been regional account manager, brand manager, and customer marketing director for Austria and Germany. R&D and materials engineering directors were recruited internally from among younger staff.

At Albus, the whole management team was changed right after privatization. The new general manager has 10 years of experience in this industry—at Albus, she started as chief of production, then became manager of the plant, before her recent promotion to general manager. The acting financial manager started in one of the companies owned by the holding company, then moved to the holding company itself, before being transferred to Albus.

*Investment Climate.* According to KPMG's due diligence ahead of the tender Merima was a market leader in detergents even before privatization. As markets opened to competition in 2001, the company significantly increased rebates and increased the credit period for its customers, which enabled the company to hold on to its dominant market share. The Nomura Information Memorandum states that: "Merima's greatest strength lies in its dominant position across most of its products. It controls 79 percent of the official detergent market and 32 percent of the household cleaning products market." During our interviews, we were told by the general manager that Henkel recognized the strengths of the local Merima brands, particularly MERIX detergent, which held 60 percent of the market share in the 1990s, but whose shares declined to 30 percent by 2003, due to imports.

Henkel Merima is still a leader, with 40 percent value of market share for powder detergents in Serbia. Pervol detergent holds 55 percent value of market share in its category; Mer Glass holds above 50 percent, softener Silan around 30 percent, and Merima Medicated baby soap 17 percent. In building and consumer adhesives, Henkel Merima is the industry pioneer and top producer with 40 percent value of market share for consumer adhesives. In January 2007, Henkel Merima started exporting to Romania and Bulgaria, which required the doubling of production of detergents in Krusevac. The company is planning to start exporting to Slovenia and Croatia by the end of 2007. Because tariffs remain high relative to transport costs, Henkel has, up to now, served neighboring markets (for example, Bulgaria) from Poland.

Albus currently is exporting to the ex-Yugoslav republics of Bosnia and Herzegovina, FYR Macedonia, and Montenegro, and has plans to expand to Croatia, Slovenia, Bulgaria, Romania, Hungary, Russia, Belarus, and Ukraine. The company had 11 percent of the export market share in 2006, and the goal is to increase this share to 20 percent in the short run.

## Policy Implications from Case Studies

We discuss the policy implications of the case study findings by offering answers to the questions presented at the beginning of this chapter that bear upon the following issues: privatization design, investment climate, managerial skills, corporate governance, private R&D, and public R&D.

### Impact of Ownership on Technology Absorption

The financial and qualitative results from the companies we studied indicate that the extent of technology absorption depends on the specific incentives of the investors. In companies bought by *domestic investors*, the motivation seems to be primarily horizontal or market seeking, with the aim of winning a substantial share of the domestic market. Domestic investors tended to repair and refurbish production assets and make targeted investments in new equipment to remove bottlenecks, and their R&D strategy includes (passive) adaptation and (active) imitation of new foreign products launched in the market.

More radical changes in product mix and manufacturing took place in companies bought by *foreign investors*, where both horizontal and vertical (cost-minimizing) objectives were relevant. The closures and switching of product lines indicate that MNEs took into account the cost conditions of neighboring plants that could serve the same export markets when rationalizing the product mix. Economies of scale appear to provide the economic rationale for this process of regional or even global specialization. This rationale determined the product mix selected and, importantly, the extent and orientation of R&D after acquisition. Generally speaking, the research capabilities of the entering MNE are so advanced that only minimal domestic R&D for absorption is carried out, in contrast to the large-scale innovation of the MNEs at their R&D sites.

There is support for these findings regarding the incentives by type of owner in the EU Impact Study. In a broader sample, it finds that companies privatized according to the 1997 law and dominated by insider owners, on average have poorer financial results and performance. Their level of sales has not increased, while average salaries have. There are no signs of significant efficiency improvements or modernization efforts in the form of investments. Gradual changes implemented by these companies have not been supported by investments, and their aging production equipment and facilities will certainly worsen their situation if no action is taken. However, those companies that changed ownership (that is, the employees' shares were bought by an investment fund or a strategic investor), show a marked advantage in performance compared to those with employee ownership.

We believe that by understanding the factors affecting what *investors* stand to gain from reducing the gap from the global technology frontier by introducing new products, machinery, and training, it is possible to provide useful policy implications. The results show clearly that country-level investment strategies of MNEs are not formulated according to relative domestic costs alone, but are contingent on decisions for nearby production facilities. Consequently, the potential for attracting strategic FDI through incentives will depend on regional and global variables that are not under the control of governments. One policy implication is that attracting FDI into certain industries needs to be considered in terms of a race against other neighboring countries. In this context, the saturation of markets owing to imports will lower the value of further investments, so there could be a tradeoff between liberalization and FDI promotion.

Since we know from the literature that concentrated ownership is important for corporate governance, and particularly for innovation and necessary risk taking, governments should facilitate FDI via a properly regulated M&A process and, if still relevant, via good case-by-case tender privatization design. Such actions will increase the probability of attracting buyers with strong incentives to make substantial absorption-related investments. The dispersed ownership resulting from mass privatization (or from the 1997 law in Serbia) has proven to be particularly problematic in post conflict-countries plagued by ethnic and social divisions, such as Bosnia and Herzegovina, Moldova, Armenia, FYR Macedonia, and Tajikistan. In such circumstances, a strategic owner, local or foreign, is a *sine qua non* condition for good corporate governance, and consequently, for technology absorption.

A proposal of this study is that in insider-dominated, properly managed companies with the potential to attract FDI through M&A, the government could facilitate consolidation of more than 51 percent of the shares together with minority shareholders, which could be attractive enough to entice a strategic investor. In addition, in M&A the government could facilitate:

- hiring of high-quality financial advisers for transactions;
- attracting a core (strategic) investor by accepting lower revenue for the sale of government shares;
- avoiding investment and employment commitments; and
- clearing past debts to the state.

## *The Effect of Corporate Governance on Absorption*

The introduction of modern corporate governance arrangements, where management is delegated control over most operating decisions by shareholders, who in turn have responsibility for monitoring and making key strategic decisions, was seen as very positive by all the managers interviewed.

A comparison of companies bought by local non strategic investors versus foreign strategic investors shows marked differences in terms of the sophistication of the arrangements and the degree of separation between ownership and control. Simplifying for the sake of clarity, we could say that domestic-owned companies exert more direct supervision and control over operating decisions, and do so by a direct relationship between the owner(s) and top management. In the case of the foreign investors, rules and norms regulate reporting lines between the MNE and the subsidiary (for example, a matrix structure), and decision making is tied to long-term planning methods that have to be agreed upon and adhered to.

Although it falls outside the scope of this report to make detailed recommendations about corporate governance, the diversity of governance arrangements between investors and acquired companies suggests that the government may want to introduce rules to ensure minimum corporate governance after acquisition. Specifically, two measures that could be considered are the adoption and disclosure of corporate governance guidelines, and requirements about independent directors. Corporate governance is very relevant to our topic since we believe that it is a necessary condition to ensure incentives for risk taking, which is a prerequisite for innovation and technology absorption.

## Management and Organizational Change

An essential part of successful FDI-driven absorption concerns the development of a competent managerial cadre with the appropriate incentives and tools. Managerial competences that are used to effect far-reaching changes in technology, workforce organization, etc., need to be developed, exercised, and rewarded. This issue was identified in each of the cases, and could be highlighted as one of the triggers for a broad corporate transformation process that increases the value added. Yet, there is no unique solution: Some companies decide to replace all pre-acquisition management, others decide to keep most of the team; in some companies, the development of younger staff is paramount, in others, the top managers are brought from outside, whether it be from the staff of the strategic investor or through head hunting from competitors. A common post-acquisition change is the introduction of a more powerful incentive structure for managers and workers.

This pattern is consistent with the EU Impact Study (p. 39), which compares the 1997 and 2001 laws on privatization. In the case of the 1997 law, only about half of the companies have changed their managers since privatization. The percentage is much higher in the case of 2001 privatizations (close to two-thirds of those giving information on the status of their directors). In the 1997 cases, most new directors came from inside the company; in the 2001 cases, the majority came from outside.

Overall, the study finds evidence that managerial capacity is rapidly developed by buyers, and this process is often based on informal learning by doing of appointed managers as they interact with strategic investors. In the case of FDI especially, we conclude that there is limited or no role for intervention by the government regarding post-acquisition managerial and organizational changes. However, this case study is restricted to large manufacturing companies, and public support could have a role to play for the healthy development of small and medium enterprises (SMEs) bought by investors with fewer resources and capabilities.

## FDI and Employment

One more dimension of reorganization concerns the workforce. In the companies that we examined, 20 percent to 50 percent of the workforce left the company after acquisition. This process has been accompanied by new hiring focused on sales and marketing, and the introduction of reward schemes to improve work incentives. For employees, the acquisitions had a mixed outcome: higher salaries and quality of employment were made possible by efficiency gains, but at the expense of a shrinking workforce. The restructuring process and the wider effort to minimize costs were synonymous with a reduction in employment levels. Absorption via plant modernization and automatization depressed demand for labor for a given output, as the capital-to-labor ratio of new machinery and equipment tends to be higher. However, the arrival of fresh resources from the investor—and later on the achievement of a break even—allowed companies to pay wages and social contributions regularly, and to offer relatively generous severance packages that compensated workers affected by technical redundancy. Furthermore, the modernization of the companies directly improved the conditions of work by creating safer environments.

These results agree with the EU Impact Study, which found that privatization has had a negative short-term impact on employment, due to adjustments implemented by the companies. The substantial decrease of employment associated with privatization under the 2001 law was accompanied by a change in the qualifications of employees. Average salaries of companies privatized according to the 2001 law have jumped by 130 percent the first year, to reach a total of 150 percent the second year. Companies privatized under the 1997 law didn't make significant changes in employment.

In transition economies, the expected impact of technology and knowledge transfer elicited by FDI can be large because of the complementary technical skills embodied in an educated workforce. Without these, the introduction of new machinery and the implementation of quality certification would be unfeasible. The case studies show the importance of in-house post-privatization training programs for employees. We already mentioned that appropriability issues can be a barrier to the development of local managerial and technical capabilities. The same problem applies to the technical skills embodied in an educated workforce: high turnover deters locally-owned companies from investing in in-house training (necessary to update existing skills), because they cannot run the risk of losing the newly-trained employees.

## FDI and Investment Climate

In contrast to other Eastern European socialist countries, Yugoslav enterprises had a tradition of exporting to Western Europe. Serbia's comparative advantage in the 1980s was fruit processing from the northern region of Vojvodina, and generic pharmaceuticals. As a consequence of the embargo imposed on the Milosevic regime in the 1990s, Serbian enterprises lost most of their export sales, and were only slowly rebuilding their network of customers when the 2001 privatization law came into effect. From an analytical perspective, this situation creates a "natural experiment," as we can observe the export profile prior to the embargo period and after acquisition. Every company starts from a position of forced autarchy, and once the external constraint ends, the management needs to decide how to serve foreign markets. Our case studies suggest that reestablishing a presence in foreign markets without an alliance, joint venture, or FDI, is a difficult undertaking. In general, the companies sold to domestic investors that we have examined (Albus, Nissal, Zorka) have not been able to increase exports in such a significant way, while their comparators (Merima, Seval, Zdravlje) are doing much better.

## FDI and R&D

In Serbian companies acquired by foreign investors, the comparative advantage for R&D lies in the adaptation of products and machinery to local conditions. For example, advanced formulas or product designs are transferred from the MNE and adapted locally, so that they can be manufactured efficiently in the acquired plant. There is also need for introducing minor marketing-led innovations, screening of competitor products, quality control, and establishing standards, among other activities. The underlying reasons for not maintaining large R&D facilities locally, as we pointed out, are that economies of scale and scope push toward a consolidation of innovation and production activities in large specialized facilities usually located in the same country as the headquarters of the MNE.

Policymakers need to be aware of the advantages and drawbacks of having manufacturing firms in which the owners have minimal incentives for innovation-seeking R&D, and which instead spend resources primarily on the task of transferring technology. On the positive side, this tends to accelerate the movement of the industrial base toward the global technology frontier, which is critical for increasing productivity in the short run, and increases the incentives for domestic rivals to upgrade. And technology absorption could be a necessary first step toward a more ambitious innovation agenda for the industrial sector.[53]

In Serbia, the large technology gap between foreign entrants and domestic consulting companies or domestic research organizations creates problems for collaboration, whether this regards consulting services or research consortia with local researchers. This weakens the effectiveness of supply-side government policies to promote technological progress, regardless of how well structured the policies are. For example, establishing a fund to encourage collaboration between research organizations and industry is unlikely to draw much interest from companies that have already received FDI and have advanced several steps on the technology ladder, because the subsidy cannot compensate for the transaction costs and delays involved.

To meet this challenge, the government can consider two courses. One is to concentrate its attention on creating incentives that support the absorption process by local industry; another is to make the deep-seated changes and substantial investments required to restructure the old R&D institutes (RDIs), so that the public R&D infrastructure can play an active role in the industrial transition from absorption to innovation.

As knowledge, commercial innovation, and R&D become a priority in ECA's advanced reformers, the industrial R&D institutes (IRDIs), inherited from the centrally planned system, have not been restructured in many ECA countries. Scarce resources spent on subsidizing IRDIs could have been used more efficiently to encourage innovation. In addition, the restructuring of IRDIs would stimulate the transition of applied R&D and laboratory workers to private enterprises. Restructuring would resolve some of the current intellectual property conflicts of interest created by the systemic moonlighting of RDI workers in private enterprises.

---

53. "Key Figures 2007 On Science, Technology And Innovation: Towards A European Knowledge Area" Monday 11 June 2007, http://ec.europa.eu/invest-in-research/pdf/kf_2007_prepub_en.pdf

# Appendixes

APPENDIX A

# Statistical Tables for Chapter 2

Table A.1. Patent Citations in ECA and Comparator Regions

| Variable | Number of Citations | Mean Application Year | Generality | Subsequent Citations |
|---|---|---|---|---|
| ECA (all) | 7.1889 | 1984.764 | 0.4657 | 14.432 |
| ECA (indigenous) | 5.7527 | 1983.984 | 0.4493 | 13.3107 |
| Emerging Asia (indigenous) | 6.8698 | 1985.798 | 0.4689 | 13.541 |
| Advanced Asia (indegenous) | 5.3601 | 1987.297 | 0.4669 | 13.4304 |
| European periphery (indigenous) | 7.452 | 1985.263 | 0.4671 | 15.1387 |
| European Core (indigenous) | 6.017 | 1985.888 | 0.4641 | 14.2001 |
| Latin America | 8.1445 | 1983.623 | 0.4129 | 11.7272 |

*Source:* Authors' calculations based on use of Bronwyn Hall's update of the NBER Patent Citation Database. Data include patents granted through 2002.

### Table A.2. Hypothesis Tests for Equality of Sample Means

| Variable | Number of Citations | Mean Application Year | Generality | Subsequent Citations |
|---|---|---|---|---|
| ECA vs. Emerging Asia | −5.161 (0.000) | −6.576 (0.000) | −1.929 (0.0537) | +0.246 (0.8054) |
| ECA vs. Advanced Asia | −0.588 (0.5563) | −17.019 (0.000) | −2.465 (0.0137) | −4.303 (0.000) |
| ECA vs. EU Periphery | −10.053 (0.000) | −6.417 (0.000) | −2.229 (0.0258) | −7.229 (0.000) |
| ECA vs. EU Core | −5.092 (0.000) | −9.880 (0.000) | −2.005 (0.0449) | −5.515 (0.000) |
| ECA vs. Latin America | −11.225 (0.000) | +1.763 (0.078) | +3.598 (0.000) | +0.348 (0.7275) |

*Note:* The test statistic is a z-statistic. A (−) sign indicates that ECA is smaller than the comparator region, a (+) sign indicates that ECA is bigger. P-values are given in parentheses. Statistically significant differences are in bold.

*Source:* Authors' calculations based on use of Bronwyn Hall's update of the NBER Patent Citation Database. Data include patents granted through 2002.

# APPENDIX B

# Regression Variables Used in Chapter 3

Table B.1. Definition of Variables Used in Regressions

| Variable Name | Definition |
|---|---|
| New Product | Dummy variable equal to 1 if the firm has introduced a new product or process. |
| Product Upgrade | Dummy variable equal to 1 if the firm upgraded an existing product or process. |
| New Technology | Dummy variable equal to 1 if the firm has introduced a new technology. |
| Upgrade | The sum of firm responses to questions on: (1) introduction of a new product or process, (2) upgrading of existing product or process, (3) achievement of new quality certification, and (4) new technology licensing agreement. The variable takes the values 0 to 4. |
| Export dummy | Dummy variable equal to 1 if the firm exports some of its output directly or indirectly. |
| Exports as a percentage of sales | Percentage of sales that are exported directly or indirectly. |
| Majority foreign owned | Dummy variable equal to 1 if percentage of firm owned by private foreign individual/company > 50%. |
| Percentage sales to MNCs | Percentage sales to multinational corporations. |
| Joint ventures with MNCs | Dummy variable equal to 1 if the firm has a joint venture with an MNC. |

*(continued)*

## Table B.1. Definition of Variables Used in Regressions (*continued*)

| Variable Name | Definition |
| --- | --- |
| Size | Based on the total number of employees in the firm: size = 1 if 0 < employees < 50; size = 2 if 50 < = employees < 250; size = 3 if employees > = 250. |
| Age | Year of the survey minus the year when the firm started operation. |
| State ownership | Dummy variable equal to 1 if percentage of firm owned by the state > 50%. |
| R&D expenditure | Log of the expenditure on research and development by the firm. |
| Foreign ownership | Percentage of the firm owned by private foreign individual/company. |
| ISO Certification | Dummy variable equal to 1 if the firm obtained a new quality accreditation (ISO 9000) in the three years |
| Web Use | Dummy variable equal to 1 if the firm uses e-mail and the Internet regularly in its interactions with clients and suppliers. |
| Training | Dummy variable equal to 1 if the firm offers training for its employees. |
| Skilled workforce | Percentage of the firm's current permanent full-time workers that are professionals (e.g., accountants, engineers, scientists). |
| University graduates | Percentage of the workforce that has a university degree or higher. |
| Use of a loan | Dummy variable equal to 1 if the firm used a loan. |
| Infrastructure Index | First principal component derived from factor analysis of (1) the average in the firm's city of the number of days with power outages or surges from the public grid, and (2) the average in the firm's city of the number of days with unavailable mainline telephone services in the year prior to the survey. |
| Governance Index | First principal component derived from factor analysis of (1300) 100% minus the percentage of firms that report that it is frequently, usually, or always true that firms in their line of business have to pay some irregular "additional payment/gifts" to get things done with regard to customs, taxes, licenses, regulations, services, (2) the percentage of firms that tend to agree, agree in most cases, or strongly agree with the statement "I am confident that the legal system will uphold my contract and property rights in business disputes," (3) the percentage of firms that tend to agree, agree in most cases, or strongly agree that the interpretations of the laws and regulations affecting the firm are consistent and predictable, and (4) 100% minus the average of the percentage of senior management time spent dealing with public officials about the application and interpretation of laws and regulations and to get or to maintain access to public services in the year prior to the survey. |

*Source:* BEEPS datasets 2002, 2005, EBRD-World Bank.

APPENDIX C

# Questionnaire for Company Interviews

Version: March 20, 2007. Used in Case Study: Technology Absorption Following Brownfield Investment and Privatization in Serbia.

## Innovation and Investment

1. **Can you tell us about the introduction of any NEW products (goods, services) post-privatization?**

   For example, we are interested in knowing:

   - If the products were developed by your enterprise, developed together with other enterprises or institutions, transferred by your enterprise group, acquired externally.
   - The costs and time taken by the development or transfer, and how this process was organized (appraisal, decision, implementation, evaluation)
   - Whether the product was new to your market, or only new to your enterprise.
   - If the products introduced were protected by legal methods (trademarks, patents, confidentiality agreements) or strategic methods (secrecy, lead time).
   - The main objectives for introducing these products (e.g., changing the product mix, expanding to new markets, environmental impact).

2. **Can you tell us about the introduction of any IMPROVED *products* (goods, services) post-privatization?**

   Again, we are interested in knowing:

   - If the improvements were carried out by your enterprise, together with other enterprises or institutions, transferred by your enterprise group, acquired externally.

- The costs and time taken by the development or transfer, and how this process was organized (appraisal, decision, implementation, evaluation)
- Whether improvements were new to your market, or only new to your enterprise.
- If improvements were protected by legal methods (trademarks, patents, confidentiality agreements) or strategic methods (secrecy, lead time).
- The main objectives for introducing these improvements (e.g., changing the product mix, expanding to new markets, environmental impact).

3. **Can you tell us about the introduction of any NEW manufacturing *processes (as opposed to "product" in the last 2 questions)* post-privatization?** For example, a process change could be a change in the layout of the production floor, a new technique to separate chemical ingredients, energy or other input-saving technologies.

   For example, we are interested in knowing:

   - If the manufacturing innovations were developed by your enterprise, developed together with other enterprises or institutions, transferred by your enterprise group, acquired externally.
   - The costs and time taken by the development or transfer, and how this process was organized (appraisal, decision, implementation, evaluation)
   - Whether the manufacturing innovations were new to the industry.
   - If the processes were protected by legal methods or strategic methods.
   - The main objectives for introducing these manufacturing innovations (e.g., more production flexibility, reducing unit costs, increasing capacity).

4. **Can you tell us about the introduction of any IMPROVED manufacturing *processes (as opposed to products above)* post-privatization?**

   Again, we are interested in knowing:

   - If improvements were developed by your enterprise, developed together with other enterprises or institutions, transferred by your enterprise group, acquired externally.
   - The costs and time taken by the development or transfer, and how this process was organized (appraisal, decision, implementation, evaluation)
   - Whether the improvements were new to the industry.
   - The main objectives for introducing these improvements (e.g., production flexibility, reducing unit costs, waste reduction, increasing capacity).

5. **Can you describe the channels of product and technology acquisition and development undertook post-privatization?**

   For example, have there been changes in the financing or organization of:

   - In-house R&D,
   - Contract R&D carried out in research institutes or other external organization
   - Acquisition of machinery, equipment and software, WHICH brought about new technology or knowledge in the firm.
   - Purchase or licensing of patents and non-patented technology, know-how, etc. from *local* sources

- Purchase or licensing of patents and non-patented technology, know-how, etc. from *foreign* sources

6. Can you explain any cooperation agreements you have with enterprises or institutes to undertake your technology acquisition and development activities?

   For example, we would like to know about:

   - Cooperation with the foreign investor which invested in your company
   - Cooperation with other foreign organizations
   - Cooperation with domestic organizations
   - Characteristics of your co-operation partners (e.g., enterprises, consultants, research institutes, suppliers, clients)
   - Their location (local/regional, in Serbia, elsewhere in Europe, other countries)

7. Can you give us details about the guaranteed investment plan signed at privatization?

   For example, has the structure (amounts, breakdown) and timing of investments corresponded to the commitments made at the time of privatization.

## Standards and Quality Certification

8. Can you describe the process of quality improvement post-privatization and whether was accompanied by certification (e.g. ISO 9000) or other methods for quality control and assurance?

9. Has the company adopted international accounting standards as provided by the International Accounting Standards Board IASB (IFRS) or US GAAP?

## Skills and Training

10. Can you discuss the formal training provided or financed by your company?

    For example, we are interested in knowing:

    1. The breakdown between training areas: management, administrative skills, customer relations, technical skills, ICT, occupational safety, basic skills,
    2. If provision is internal or external, and whether foreign providers were used
    3. The percent of employees that received formal training in the last 12 months and the average number of training hours per person
    4. If the training budget increased post-privatization and whether the fraction spent in-house and on external providers has changed
    5. Whether training is provided to update skills because of existing and planned investments in new machinery and equipment

11. Can you tell us about the post-privatization status of information and communications technology (ICT) skills of your current employees?

For example, the percentages of employees that have acquired:

- Basic computer skills,
- Good working knowledge of standard office software
- Facility with industry specific software
- Knowledge of hardware and maintenance
- Advanced knowledge of programming and application or system architecture,

12. **Can you tell us about changes to human resources (HR) policy and development planning?**

    For example, we are interested in knowing whether there have been:

    - Changes to the recruitment policy
    - Performance-oriented incentive schemes for employees
    - Regular performance reviews, staff meetings and career planning
    - Definition and agreement of personnel goals

13. **Can you tell us about post-privatization developments in the labor market relevant to your industry?**

    For example, we are interested in knowing about:

    - Any difficulties in recruiting personnel with the skills needed for current production processes or in R&D
    - Competition for employees—for example, whether any employees left to join other firms in the same industry during the last 12 months
    - The trends in wages for personnel in management, production and research

## Business Practices and Management

14. **Can you tell us about major changes in your business structure and systems post-privatization?**

    For example, have any of the following been implemented:

    - Advanced management techniques (e.g. IT-based management information systems)
    - Introduction of cross-functional teams
    - Outsourcing of major business functions, disposal of non-core activities
    - Rationalization of distribution channels (e.g., bulk and non-bulk shipments)
    - Changes in marketing strategies (e.g. packaging or presentational changes to target new markets, advertising spending)
    - Type of reporting lines and reporting frequency

15. **Can you describe any changes in senior management?**

    For example:

    - How many executive directors and members of the supervisory and management boards changed post-privatization

- What is the background of new directors or board members (previous experience in the country and abroad, type of education—university, MBA)
- Changes in the functions of the board or different departments/divisions changed

16. **What if any changes are there in the relationship with suppliers of raw material water, energy and packaging?**

    For example, we are interested in knowing in knowing if:

    - New suppliers replaced pre-privatization suppliers
    - New suppliers are domestic or foreign, and if they belong to the company's global supply chain
    - Contract terms with pre-privatization suppliers changed (e.g. price, quality, timing)
    - Cost implications of new procurement and delivery arrangements

## Competition and Exports

17. **How have competition and exports changed since privatization?**

    For example, we are interested in information about:

    - The number and size of competitors in the national market for your main product lines or services. Have there been changes post-privatization?
    - Competition from imports in your main product lines or services, and the country of origin of these goods and services. Have there been changes post-privatization?
    - Any new markets you are exporting to and what role your enterprise played in these developments

## Government Support

18. **Can you discuss the types of public financial support your company received pre- and post-privatization?**

    For example, can you provide information about:

    - Any funding from the Development Fund as a private company? Any funding as a socially owned company before privatization?
    - Overdue tax arrears and arrears to EPS
    - Payments from the Transition Fund to any employees laid-off from the company before and after privatization
    - Any grants, loans or tax deductions you receive for investment in R&D?

# APPENDIX D

# Correlates of ICT and Quality Certification[54]

The regression results in Chapter 3 showed that proxies for the use of information and communications technologies (ICT), and the adoption of global standards and technical regulations by firms in ECA countries, are positively and strongly associated with all the measures of technology absorption. These findings suggest that ICT use and the adoption of global standards and technical regulations can be seen as important channels of technology absorption, (although the empirical literature also suggests that they can have a direct impact on productivity). This Appendix examines how far the investment climate can systematically explain the gap that exists among firms in the adoption of international standards and the use of ICT. We exploit the cross-country firm-level data in the BEEPS surveys of 2002 and 2005 to provide evidence on the investment climate constraints faced by firms, and, as in chapter 3, we employ Internet use and ISO certification as proxies for firms' use of ICT, and their adoption of global standards and technical regulations, respectively.

The adoption of global industry standards and technical regulations by firms is shown to be among the most important ways of introducing product and process technology upgrading, and of increasing productivity of firms (Corbett, Montes-Sancho, and Kirsch 2005). Also, Blind, Temple, Swann and Williams (2005) found that over 60 percent of product and process innovators in the United Kingdom used technical standards as a source of information, which was twice the share of companies that cited universities or research laboratories as a source of information. As such, standards embody information on the state of the art of a particular technology available to any firm, at least in principle.

---

54. See Correa, Fernandes, and Uregian (2008, Forthcoming).

Standards play a particularly central role as channels for knowledge diffusion in industries where products and processes supplied by various providers must interact with one another. ISO certification measures a firm's adoption of international industry standards and technical regulations, and thus is a channel for codified technology absorption. Specifically, ISO 9000 norms are a set of international standards and guidelines that serve as the basis for establishing quality management systems at manufacturing and services firms (ISO 1998). ISO 9000 certification is awarded to quality processes within a firm and requires a detailed review and documentation of the firm's production processes to be in accordance with ISO quality system requirements (Guler, Guillen, and Macpherson 2002). For firms in developing countries, the adoption of international standards such as ISO certification often also creates additional productivity benefits beyond the transfer of new productivity-enhancing knowledge embodied in standards by facilitating entry into global supply chains. Due to economies of scale, this further facilitates the transfer of technical and organizational knowledge from technologically advanced buyers—usually multinational corporations (MNCs)—in value chains to local firms (Arora and Aundi 1999, and Humphrey and Schmitz 2000). Thus, ISO certification can also act as an indicator of whether firms are suppliers to MNCs. Since the products and processes involved in global value chains tend to have a higher technological content than those outside those value chains, ISO certification should definitely be a channel for higher technology absorption by the firm.

Internet use by firms is used as a proxy for the use of ICT in business operations, and thus as a channel for technology absorption. A very large body of empirical literature documents a strong positive relationship between ICT and productivity *at both the aggregate and firm levels,* particularly in the United States during the second half of the 1990s (Cohen, Garibaldi, and Scarpetta 2004; Stiroh 2002; Jorgenson 2001). The jump in productivity of the U.S. service sector during this period, epitomized by the revolution in retailing brought about by Wal-Mart via its substantial investment in ICT-based methods, subsequently spread to other manufacturing industries that rely on ICT. As such, ICT is considered to be the preeminent "general purpose technology" of the past 20 years, as it has driven economy wide growth by spreading over a range of sectors and prompting them to innovate further. Technological progress in these sectors, in turn, creates incentives for further advances in the ICT sector, thus setting up a positive, self-sustained virtuous cycle (Bresnahan and Trajtenberg 1995; Helpman and Trajtenberg 1996). According to recent studies by Piatkowski and van Ark (2005, 2006), ICT also had a positive impact on economic growth in many Central European Economies (CEE) in the ECA region during this period: ICT investment contributed between 0.5 percentage points and 0.7 percentage points of GDP growth during the period from 1995 to 2003 in Poland, Slovenia, the Slovak Republic, the Czech Republic, and Hungary—slightly surpassing its contribution in the EU-15 (though remaining smaller than its contribution in the United States by almost 0.9 percentage points). A 2007 World Bank report on productivity in the ECA region shows that productivity gains in ICT-producing or ICT-using manufacturing and service industries substantially surpassed gains in non-ICT industries in the CEE during this period. And interestingly, for all of these countries, growth was at least as high in manufacturing as it was in service industries.

## ISO Certification and Web Use in ECA

As in Chapter 3, the dataset used for this Appendix is based on the BEEPS surveys from 2002 and 2005 collected by the World Bank in 28 countries in the ECA region. All countries are represented in both surveys, with 6,667 firms surveyed in 2002, and 9,655 in 2005, and we have also composed a panel of 2,892 firms that were surveyed in both 2002 and 2005. The surveys cover a range of manufacturing and service sectors, and are designed to be representative of the universe of firms according to sector and location within each country. However, we should emphasize that the 2002 and the panel samples have a larger share of firms in the service sectors (61.4 percent and 62 percent, respectively), while the 2005 sample is split evenly between manufacturing and services (49.6 percent and 50.4 percent, respectively).

Figure D.1 shows the shares of firms with ISO certification and Internet use in 2002 and 2005, respectively, by country/subregion in Europe and Central Asia. ISO certification across the ECA firms surveyed fell slightly on average, from 13.6 percent to 12.5 percent, rising only in South-eastern Europe (SEE)[55] and Turkey, while remaining stable in all other parts of the region except Russia, where the share of ISO-certified firms in 2005 was half (9.3 percent) of that in 2002 (18.5 percent). In contrast, the share of firms with Internet use across ECA in 2005 (67.4 percent) was higher than in 2002 (58.2 percent), with the largest difference in the countries/subregions with the lowest Internet use in 2002 (Turkey, and CIS outside Russia and Ukraine). There is still substantial heterogeneity across ECA, with the "other CIS" still being the subregion with the lowest Internet usage, with just under 50 percent of firms employing ICT; but the gap relative to EU-8 members[56], which have the highest penetration (83.5 percent), has been reduced from 2002 (39 percent) to 2005 (34 percent).[57]

Table D.1 shows that there is substantial heterogeneity in ISO certification and Internet use across sectors, with firms in the manufacturing sectors being significantly more likely to use the Internet in their operations and to have ISO certification than firms in other sectors. Firms in the mining/quarrying and construction sectors are significantly more likely than the rest of the sample to have adopted ISO standards, but have an average score for Internet usage.[58] In contrast, the share of firms in the service sectors that absorbed either of these technologies was significantly lower (although there were some exceptions, such as real estate), even though ICT, in particular, has led to substantial productivity gains in the services sector in the United States and Europe. This is consistent with

---

55. Southeastern Europe (SEE) includes Albania, Bosnia Herzegovina, Serbia and Montenegro, the Former Yugoslav Republic of Macedonia, Bulgaria, Romania, Croatia.

56. EU-8 countries are the eight ECA countries that joined the European Union on May 1st 2004: the Czech Republic, Estonia, Hungary, Latvia, Lithuania, Poland, the Slovak Republic, and Slovenia.

57. Note that for any given ECA country, the changes in shares of ISO-certified firms between 2002 and 2005 are dependent on the use of the BEEPS samples, and as such they may differ from the numbers of ISO-certified firms from the official ISO statistics.

58. The fact that ISO certification is sometimes mandatory for public sector projects could explain the large number of certified construction firms. Also, mining firms are on average much larger than manufacturing firms and thus find it much easier to become ISO certified.

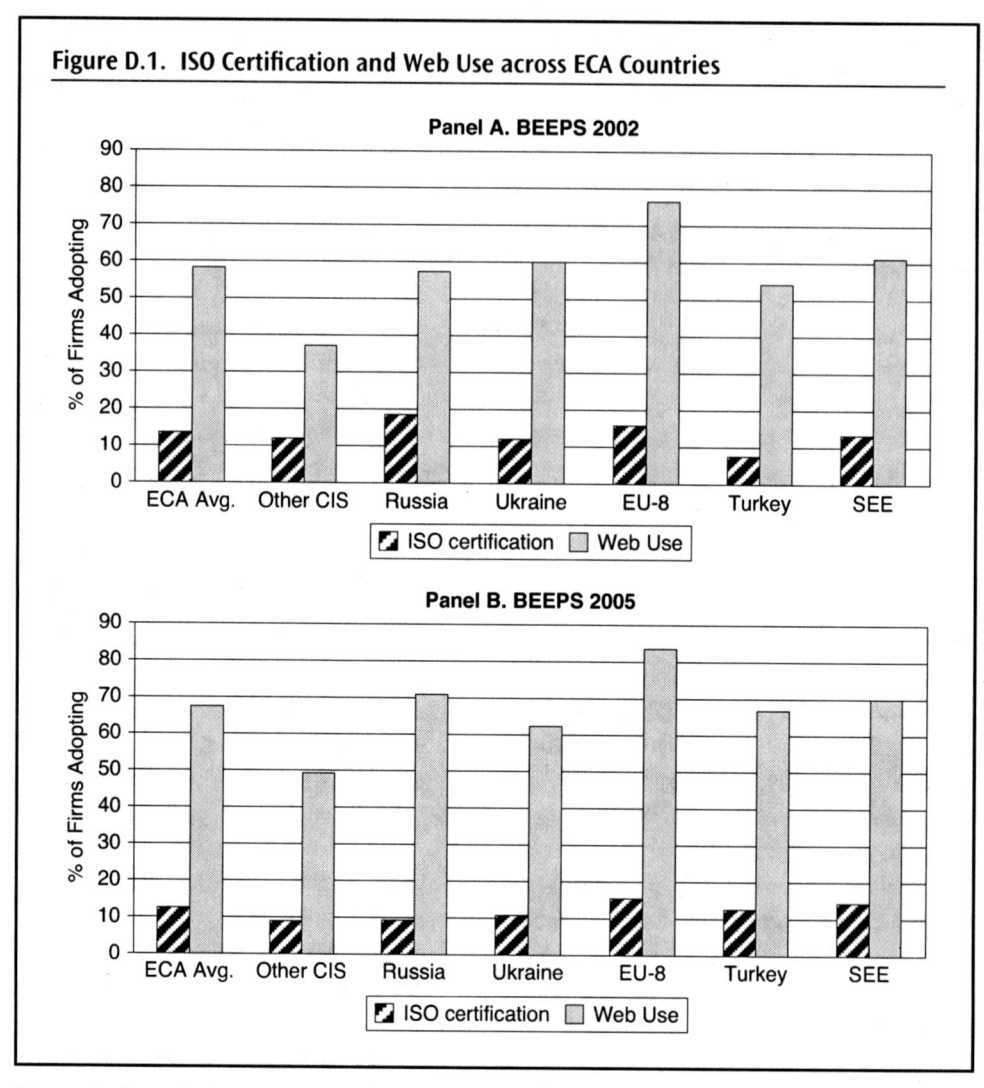

Figure D.1. ISO Certification and Web Use across ECA Countries

**Panel A. BEEPS 2002**

[Bar chart showing % of Firms Adopting ISO certification and Web Use across ECA Avg., Other CIS, Russia, Ukraine, EU-8, Turkey, SEE]

**Panel B. BEEPS 2005**

[Bar chart showing % of Firms Adopting ISO certification and Web Use across ECA Avg., Other CIS, Russia, Ukraine, EU-8, Turkey, SEE]

*Source:* Business Environment Enterprise Performance Surveys 2002, 2005.

global experience where, despite recent growth in ISO certification in the services sector, certification is more prevalent in manufacturing industries, where quality is very important for export competitiveness.[59]

Table D.2 describes variations in ISO certification and Web use across different firm characteristics. Small (1–49 employees) and medium-sized enterprises (50 to 249 employees) exhibit significantly lower shares of ISO certification and Internet use than do large firms. This finding is consistent with the literature and experience in the OECD (Cohen

---

59. Neumayer and Perkins (2005) confirm empirically the positive relationship between the share of the manufacturing sector in the economy and ISO 9000 diffusion rates.

Table D.1. ISO Certification and Web Use across Sectors

|  | ISO Certification | | Web Use | |
| --- | --- | --- | --- | --- |
|  | BEEPS 2002 Sample | BEEPS 2005 Sample | BEEPS 2002 Sample | BEEPS 2005 Sample |
| *Average across all subsectors* | 13.6 | 12.5 | 58.2 | 67.4 |
| Mining and quarrying energy-related | 36.7** | 14.3 | 70.0 | 75.0 |
| Mining and quarrying not energy-related | 21.3 | 19.4 | 62.5 | 70.1 |
| Food beverages and tobacco | 23.6*** | 16.2*** | 54.1* | 57.0*** |
| Textiles | 17.4 | 7.9*** | 58.4 | 63.0** |
| Leather | 9.7 | 8.3 | 58.1 | 54.2* |
| Wood | 23.8* | 7.1* | 50.0 | 60.0 |
| Pulp and paper | 11.2 | 13.8 | 81.0*** | 84.1*** |
| Petroleum | 8.3 | 16.7 | 91.7*** | 100.0*** |
| Chemicals | 26.6** | 36.6*** | 78.8*** | 88.1*** |
| Rubber and plastics | 24.5* | 23.3** | 62.3*** | 88.9*** |
| Nonmetallic minerals | 17.2 | 20.3** | 54.0 | 64.2 |
| Metals | 24.4*** | 19.0*** | 65.2* | 77.2*** |
| Machinery and equipment | 33.8*** | 24.9*** | 79.9*** | 82.1*** |
| Electrical and optical equipment | 34.9*** | 30.4*** | 81.1*** | 88.7*** |
| Transport equipment | 42.3*** | 42.6*** | 86.8*** | 87.2*** |
| Other manufacturing | 20.3* | 11.6 | 64.9 | 70.7 |
| *Total Manufacturing* | | | | |
| Construction | 17.4*** | 15.9*** | 56.7 | 69.9* |
| Wholesale and retail trade | 8.7*** | 8.0*** | 50.0*** | 62.3*** |
| Hotels and restaurants | 6.2*** | 7.1*** | 42.2*** | 53.0*** |
| Transport, storage, and communications | 10.0*** | 11.0 | 71.8*** | 78.5*** |
| Real estate and business activities | 10.9** | 9.2*** | 71.4*** | 79.5*** |
| Other services | 5.1*** | 4.5*** | 50.1*** | 53.7*** |

*Source:* BEEPS 2002 and BEEPS 2005 surveys.
*Notes:* Values are in percentage. ***, **, and * indicates statistical difference from the rest of the sample at 1%, 5%, and 10% confidence levels, respectively.

and Klepper 1996). The relationship between ISO certification or Internet use and firm age is also significant, but less pronounced. Table D.3 shows the association between firm ownership, trade integration, and ISO certification or Internet use. In accordance with the literature summarized in Chapter 3, which finds a positive association between knowledge transfer and FDI and trade, we find in our sample that a larger share of fully foreign-owned firms, joint ventures, and exporters use the Internet in their operations and are ISO certified, in comparison with domestic firms.

## Table D.2. ISO Certification and Web Use, Firm Size, and Age

|  | ISO Certification | | Web Use | |
|---|---|---|---|---|
|  | BEEPS 2002 Sample | BEEPS 2005 Sample | BEEPS 2002 Sample | BEEPS 2005 Sample |
| *Size* | | | | |
| Small | 8.5*** | 7.8*** | 50.0*** | 59.9*** |
| Medium | 20.6*** | 20.7*** | 70.3*** | 82.8*** |
| Large | 28.8*** | 30.0*** | 81.8*** | 91.1*** |
| *Age* | | | | |
| Quartile I | 11.1*** | 7.1*** | 49.5*** | 57.8*** |
| Quartile II | 11.5*** | 11.1** | 53.8*** | 64.2*** |
| Quartile III | 14.6 | 12.4 | 65.5*** | 70.4*** |
| Quartile IV | 16.6*** | 17.7*** | 63.0*** | 74.9*** |
| ECA Average | 13.6 | 12.5 | 58.2 | 67.4 |

*Notes:* Values are in percentage. ***, **, and * indicates statistical difference from the rest of the sample at 1%, 5%, and 10% confidence levels, respectively. For 2002: Quartile I (0–4 yrs; mean = 3.6); Quartile II (5–8 yrs; mean = 6.4); Quartile III (9–11 yrs; mean = 10); Quartile IV (11+, mean = 34.7) For 2005: Quartile I (0–3 yrs; mean = 1.9); Quartile II (4–6 yrs; mean = 5); Quartile III (7–10 yrs; mean = 8.4); Quartile IV (12+, mean = 30).

*Source:* BEEPS 2002 and BEEPS 2005 surveys.

## Table D.3. ISO Certification and Web Use, Firm Ownership, and Trade Integration

|  | ISO Certification | | Web Use | |
|---|---|---|---|---|
|  | BEEPS 2002 Sample | BEEPS 2005 Sample | BEEPS 2002 Sample | BEEPS 2005 Sample |
| Recently Privatized | 20.2*** | 19.2*** | 54.4** | 66.2 |
| Private | 11.8*** | 10.7*** | 58.2 | 66.8*** |
| Foreign | 23.4*** | 19.6*** | 88.4*** | 89.2*** |
| Joint Venture | 22.7*** | 21.9*** | 82.1*** | 86.2*** |
| Exporter | 25.1*** | 22.9*** | 83.4*** | 90.1*** |
| ECA Average | 13.6 | 12.5 | 58.2 | 67.4 |

*Notes:* Values are in percentage. ***, **, and * indicates statistical difference from the rest of the sample at 1%, 5%, and 10% confidence levels, respectively.

*Source:* BEEPS 2002 and BEEPS 2005 surveys.

## Determinants of ISO Certification and Use

### Conceptual Framework

To guide our empirical analysis of the main determinants of ISO certification and use by firms in the ECA region, we outline a framework that assumes firms make a rational cost/benefit assessment of each potential technology absorption-related investment—drawing on a wide literature to lay out the microeconomic factors affect that assessment. Our framework pools these factors into three groups. The first group, "market incentives," refers to four core aspects of a competitive market economy Private ownership and control establish profit maximization as the "modus operandi" of the firm and drive investment in the most productive technologies with ensuing productivity gains (Brown, Earle, and Telegdy 2007). An important complementary condition is strong (product market) competition from domestic and foreign rivals or demand from consumers that create the incentives for incumbents to engage in investments in more productive technologies, rather than engaging in rent-seeking activities (Baumol 1990; Aghion and Schankerman 1998, 2003). Consistent with that vision, Comin and Hobijn (2004) find that the degree of openness to trade is one of the most important determinants of the speed at which a country adopts more advanced technologies because it introduces the pressures of foreign competition on incumbents, thereby reducing their payoff from lobbying the government to deter the adoption of new technologies. A very large body of literature documents the critical importance of sound legal institutions that guarantee adequate *protection of property rights and contract enforcement* for the growth and development of market economies (North 1990; Nicolletti and Scarpetta 2003; Acemoglu, Johnson, and Robinson 2004). Acemoglu, Atràs, and Helpman (2007) show, in particular, how greater contractual incompleteness leads to the adoption of less advanced technologies, and how the impact of contractual incompleteness may generate sizable productivity differences across countries with different contracting institutions. Finally, investors need to be confident that they will be able to reap the benefits from their absorption-related investments, because during the initial stages of the absorption process, investments in new technology may generate productivity losses related to capital and skills specificities, adjustment costs, and learning by doing (Pavlova 2001; World Bank 2004). Adjustment and learning costs are proportional to the technology absorption gap. In this sense, a *predictable economic and regulatory policy environment*, which reduces the risk of expropriation, is the fourth key aspect of a market economy affecting a firm's decision to become ISO certified or to use the Internet.

The second group of determinants of the firm's ISO certification and Internet use decisions involves *complementary input markets*. Firms' access to complementary inputs affects the adjustment costs, and therefore the overall payoff decision to upgrade technology. Firms' access to complementary inputs differs substantially across countries, due to the fact that the availability of such inputs (e.g., skills) varies with the stage of development at which a country is (Zeira 1998). Three types of complementary inputs are particularly relevant. There is consensus in the growth literature on the importance of *human capital,* but there is also substantial empirical evidence documenting how weak human capital can delay a firm's modern technology absorption activities, in part due to the high costs of having to retrain workers to use new technologies (Berdugo, Sadik, and Sussman 2005; Alesina and Zeira 2006; Navaretti, Soloaga, and Takacs 1998). The lack of quality managerial capacity or "entrepreneurship" can also constrain the firm's investment in advanced technology, as it reduces the firm's information on the most appropriate technology, and can increase its adjustment costs (Cohen and Levinthal 1989).

*Access to finance* is a second key complementary input required for technology absorption investments by firms. Levine (2003) surveys the vast empirical literature on the role of financial development for growth, and concludes from country-, industry- and firm-level studies that there is causal evidence.[60] The importance of *physical infrastructure* for firm productivity and economic growth has been emphasized widely, and finds strong support from empirical work in OECD and developing economies (Roever and Waverman 2001; Esfahani and Ramirez 2003; World Bank 2005).

The third group of determinants of the firm's ISO certification and Internet use decisions relates to the firm's access to international knowledge, whether that knowledge be transferred via FDI or through its participation in export markets. Chapter 3 extensively describes the literature related to this group of determinants.

## Main Findings

We now discuss the findings from regressions that estimate the determinants of ISO certification and Internet use across firms in ECA, based on the framework described above. Box 1 describes our identification strategy in detail. In the regressions, we consider as variables proxying for market incentives: the type of ownership of the firm (private or recently privatized); the degree of control by majority shareholders; measures of competition; and a governance index based on measures of the quality of contract enforcement, the predictability of government regulations, and the bureaucratic burden felt by firms. We consider as variables proxying for input markets: firm size; measures of managerial capacity and worker skills; R&D efforts; access to finance; and infrastructure quality. Finally, we consider as variables proxying for access to knowledge: foreign ownership (full or through joint ventures); and the firm's participation in export markets. All variables are defined in Table D.4.

We should note a key methodological issue at this point. Our main results on the determinants of ISO certifications and Internet use are shown in Table D.5 using alternatively the BEEPS 2002 sample and the BEEPS 2005 sample; and in Table D.7 using the BEEPS panel sample. The results in Table D.5 are based on cross-sections of firms in ECA countries in 2002 and 2005, and therefore can identify systematic and robust correlations between input markets, market incentives, and access to international knowledge on the one hand, and ISO certification or Internet use on the other, but they cannot identify causality. For example, firms with better managers may be more likely to adopt technology, but also to hire a larger share of professionals, and to participate in export markets. Thus, the effect of the share of professionals and of export shares may simply reflect omitted managerial or other firm characteristics. Our strategy to address this problem is twofold. On the one hand, we include many important firm-level controls in the ISO certification and Internet use regressions to minimize the possibility that our results are

---

60. See Gatti and Love (2006) for the links between firm performance and access to finance in Bulgaria.

> **Box D.1: Identification Strategy**
>
> Our empirical strategy considers profit-maximizing firms deciding whether or not to be ISO certified and to use the Web. A firm decides to be ISO certified or use the Internet if the benefits from this decision are larger than its costs.[1] Let $\pi_{ijc}^{ISO}$ and $\pi_{ijc}^{Web}$ be the net benefits (benefits minus costs) for firm $i$ in sector $j$ in country $c$. If ISO and Web are dummy variables that equal 1 if firm $i$ is ISO certified or uses the Web, respectively, we assume that:
>
> $$ISO_{ijc} = \begin{cases} 1 \text{ if } \pi_{ijc}^{ISO} > 0 \\ 0 \text{ otherwise.} \end{cases}$$
>
> $$Web_{ijc} = \begin{cases} 1 \text{ if } \pi_{ijc}^{Web} > 0 \\ 0 \text{ otherwise.} \end{cases}$$
>
> While the net benefits $\pi_{ijc}^{ISO}$ and $\pi_{ijc}^{Web}$ are unobserved, we assume that they are a function of variables proxying for market incentives $Inc_{ijc}$, for input markets $Inp_{ijc}$, and for access to international knowledge $Kno_{ijc}$, . We assume a linear form:
>
> $\pi_{ijc}^{ISO} = \alpha_1 Inc_{ijc} + \beta_1 Inp_{ijc} + \gamma_1 Kno_{ijc} + l_j + \theta_1 GDPpc_c + \varepsilon_{ijc}$
>
> $\pi_{ijc}^{Web} = \alpha_2 Inc_{ijc} + \beta_2 Inc_{ijc} + \gamma_2 Kno_{ijc} + l_j + \theta_2 GDPpc_c + \mu_{ijc}$, where $l_j$ are sectoral fixed effects, $GDPpc_c$ is the country GDP per capita, and $\varepsilon_{ijc}$ and $\mu_{ijc}$ capture unobserved firm characteristics influencing the ISO certification or Web use decisions. Then, the probability that firm i is ISO certified or uses the Web is given by:
>
> $Pr(ISO_{ijc} = 1) = Pr(\varepsilon_{ijc} > -\alpha_1 Inc_{ijc} - \beta_1 Inp_{ijc} - \gamma_1 Kno_{ijc} - l_j - \theta_1 GDPpc_c)$
>
> $Pr(Web_{ijc} = 1) = Pr(\mu_{ijc} > -\alpha_2 Inc_{ijc} - \beta_2 Inp_{ijc} - \gamma_2 Kno_{ijc} - l_j - \theta_2 GDPpc_c)$.
>
> Assuming that the residuals $\varepsilon_{ijc}$ and $\mu_{ijc}$ are normally distributed, we can estimate each of these equations by maximum likelihood (probit). It is important to control for sectoral dummies in the regressions since sectoral differences in production technology, product demand, or competition may affect the incentives of firms to absorb new technology (Cohen and Levinthal 1989). While we are unable to include country fixed effects in our regressions due to their collinearity with the infrastructure index and the governance index (that are location specific), we do include the country's GDP per capita in order to control for country heterogeneity in ISO certification and Web use that is not captured by market incentives, input markets, or access to international knowledge.

driven by omitted variables. However, we must acknowledge that our findings based on the cross-sections of firms in 2002 and 2005 could be partly driven by unobservable firm characteristics. On the other hand, we also estimate our main specifications based on the BEEPS panel sample. In this case, we are able to control for unobserved firm invariant characteristics that may drive the determinants as well as the ISO certification and Internet use decisions. Unfortunately, this approach suffers from two drawbacks: (1) the panel dimension of the data is small (only two years of data per firm); and (2) the sample size in the panel is much smaller than in the cross-sections. The results from the panel regressions are shown in Table D.7.

## Input Markets

We first discuss the findings on the importance of *input markets* in Table D.5. Larger firms have a significantly higher propensity to be ISO certified and to use the Internet. The

## Table D.4. Variable Definitions

| Variable Name | Definition |
| --- | --- |
| ISO certification | Dummy variable equal to 1 if the firm obtained a new quality accreditation (ISO 9000) in the three years prior to the survey |
| Web use | Dummy variable equal to 1 if the firm uses e-mail and the Internet regularly in its interactions with clients and suppliers |
| Age | Year of the survey minus the year when the firm started operation |
| Size categories | Based on the total number of permanent workers employed by the firm |
| Manager with college education or more | Dummy variable equal to 1 if the firm's general manager's highest level of education is a university degree or a higher university (post-graduate) degree |
| Manager age | Age of the firm's general manager |
| Share of professionals | Percentage of the firm's current permanent full-time workers that are professionals (e.g., accountants, engineers, scientists) |
| R&D intensity | Share in total firm sales of R&D expenditures (including wages and salaries of R&D personnel, materials, R&D-related education and training costs) |
| Access to finance dummy | Dummy variable equal to 1 if the firm has a bank loan or overdraft |
| Infrastructure index | First principal component derived from factor analysis of (1) the average in the firm's city of the number of days with power outages or surges from the public grid, and (2) the average in the firm's city of the number of days with unavailable mainline telephone service in the year prior to the survey |
| Dummy for recently privatized firm | Dummy variable equal to 1 if the firm was established as the privatization of a state-owned firm |
| Dummy for private firm (from origin) | Dummy variable equal to 1 if the firm is originally private from the time of startup |
| Ownership share of largest shareholder | Percentage of the firm's equity owned by the largest shareholder |
| Dummy for market share equal to less | Dummy variable equal to 1 if the firm's percentage of the total market sales is less than 5% (available only in BEEPS 2002) |
| Price-cost margin (%) | Margin by which the firm's sales price for its main product line or main line of services in the domestic market exceeds its operating costs (i.e. materials inputs costs plus wages costs but not overheads and depreciation) |
| Dummy for pressure to innovate from competitors being important | Dummy variable equal to 1 if the firm ranks pressure from domestic or foreign competitors as being fairly important or very important for the firm's decisions about developing new products or services and markets |
| Dummy for pressure to innovate from consumers being important | Dummy variable equal to 1 if the firm ranks pressure from customers as being fairly important or very important for the firm's decisions about developing new products or services and markets |

*(continued)*

## Table D.4. Variable Definitions (*continued*)

| Variable Name | Definition |
|---|---|
| Elasticity of demand faced by firm | If the firm were to raise the prices of its main product line or main line of services 10% above their current level in the domestic market (after allowing for any inflation), what would happen assuming that the firm's competitors maintained their current prices: (1) customers would continue to buy from the firm in the same quantities as now, (2) customers would continue to buy from the firm but at slightly lower quantities, (3) customers would continue to buy from the firm but at much lower quantities, (4) many customers would buy from the firm's competitors instead |
| Number of competitors | Number of competitors faced by the firm in its main product in the domestic market: none, 1–3, 4 or more |
| Elasticity of demand faced by firm | If the firm were to raise the prices of its main product line or main line of services 10% above their current level in the domestic market (after allowing for any inflation), what would happen assuming that the firm's competitors maintained their current prices: (1) customers would continue to buy from the firm in the same quantities as now, (2) customers would continue to buy from the firm but in slightly lower quantities, (3) customers would continue to buy from the firm but in much lower quantities, (4) many customers would buy from the firm's competitors instead |
| Number of competitors | Number of competitors faced by the firm in its main product in the domestic market: none, 1–3, or 4 or more |
| Governance index | First principal component derived from factor analysis of (1) 100% minus the percentage of firms in the firm's city that report that it is frequently, usually, or always true that firms in their line of business have to pay some irregular "additional payment/gifts" to get things done with regard to customs, taxes, licenses, regulations, services, (2) the percentage of firms in the firm's city that tend to agree, agree in most cases, or strongly agree with the statement "I am confident that the legal system will uphold my contract and property rights in business disputes," (3) the percentage of firms in the firm's city that tend to agree, agree in most cases, or strongly agree that the interpretations of the laws and regulations affecting the firm are consistent and predictable, and (4) 100% minus the average in the firm's city of the percentage of senior management time spent dealing with public officials about the application and interpretation of laws and regulations, and to get or to maintain access to public services in the year prior to the survey |
| Foreign owned | Dummy variable equal to 1 if 100% of the firm's capital is owned by foreigners |
| Joint venture | Dummy variable equal to 1 if more than 0% but less than 100% of the firm's capital is owned by foreigners |
| Export share | Dummy variable equal to 1 if the firm exports some of its output directly or indirectly |
| GDP per capita (log) | Values in constant 2000 USD for the year 1995 (Source: World Development Indicators) |

*Note:* The source is the BEEPS 2002 or the BEEPS 2005, unless otherwise stated.

## Table D.5. Determinants of ISO Certification and Web Use—Cross-Sectional Regressions

| | Dependent Variable Is | | | |
|---|---|---|---|---|
| | ISO Certification Dummy | | Web Use Dummy | |
| | BEEPS 2002 Sample (1) | BEEPS 2005 Sample (2) | BEEPS 2002 Sample (3) | BEEPS 2005 Sample (4) |
| **Input markets** | | | | |
| 50–249 Workers | 0.066 [0.014]*** | 0.082 [0.012]*** | 0.162 [0.019]*** | 0.195 [0.012]*** |
| More than 250 workers | 0.102 [0.019]*** | 0.161 [0.020]*** | 0.252 [0.020]*** | 0.243 [0.012]*** |
| Manager with college education dummy | 0.037 [0.010]*** | | 0.234 [0.018]*** | |
| Manager age | 0.001 [0.000]*** | | −0.003 [0.001]*** | |
| Share of professionals | 0.065 [0.020]*** | 0.068 [0.015]*** | 0.377 [0.039]*** | 0.343 [0.030]*** |
| R&D intensity | 0.162 [0.067]** | 0.575 [0.163]*** | 0.424 [0.208]** | 1.575 [0.695]** |
| Access to finance dummy | 0.046 [0.009]*** | 0.042 [0.007]*** | 0.117 [0.016]*** | 0.156 [0.011]*** |
| Infrastructure index | −0.016 [0.006]** | −0.005 [0.004] | 0.082 [0.011]*** | 0.079 [0.007]*** |
| **Market incentives** | | | | |
| Recently privatized firm dummy | 0.02 [0.017] | −0.028 [0.012]** | 0.001 [0.030] | −0.061 [0.030]** |
| Private firm from origin dummy | 0.01 [0.014] | −0.04 [0.015]*** | 0.125 [0.027]*** | 0.006 [0.025] |
| Ownership share of largest shareholder | −0.001 [0.015] | −0.033 [0.012]*** | −0.069 [0.028]** | −0.138 [0.021]*** |
| Dummy for market share less than 5% | −0.055 [0.010]*** | | −0.135 [0.017]*** | |
| Price-cost margin (%) | | 0.005 [0.024] | | 0.005 [0.046] |
| Dummy for pressure to innovate from competitors being important | −0.001 [0.011] | 0.013 [0.008] | 0.003 [0.018] | 0.041 [0.014]*** |
| Dummy for pressure to innovate from consumers being important | 0.019 [0.010]* | 0.021 [0.008]** | 0.044 [0.018]** | 0.023 [0.014]* |
| Governance index | −0.009 [0.003]*** | −0.006 [0.003]** | −0.013 [0.005]** | −0.04 [0.005]*** |
| **Access to international knowledge** | | | | |
| 100% Foreign ownership dummy | 0.073 [0.023]*** | 0.049 [0.020]** | 0.277 [0.025]*** | 0.171 [0.022]*** |

*(continued)*

Table D.5. Determinants of ISO Certification and Web Use—Cross-Sectional Regressions (*continued*)

| | Dependent Variable Is | | | |
|---|---|---|---|---|
| | ISO Certification Dummy | | Web Use Dummy | |
| | BEEPS 2002 Sample (1) | BEEPS 2005 Sample (2) | BEEPS 2002 Sample (3) | BEEPS 2005 Sample (4) |
| JV dummy (Foreign ownership <100%) | 0.049 [0.017]*** | 0.032 [0.015]** | 0.195 [0.024]*** | 0.134 [0.021]*** |
| Export share | 0.082 [0.018]*** | 0.069 [0.015]*** | 0.296 [0.046]*** | 0.379 [0.042]*** |
| Observations | 5589 | 7968 | 5625 | 7965 |

*Notes:* Marginal effects at mean values from probit regressions shown. Robust standard errors in parenthesis. ***, **, and * indicates statistical significance at 1%, 5%, and 10% confidence levels, respectively. The regressions include also sectoral dummies and GDP per capita. Higher values of the infrastructure (governance) index indicate better infrastructure (governance).

marginal effects are particularly large for Internet use, and suggest that within a sector, firms with more than 250 workers are about 25 percent more likely to use the Internet than firms with less than 50 workers. These findings are consistent with two related arguments. First, as mentioned earlier, large firms benefit from economies of scale in the adaptation and development of new technology, and have more capacity to finance technology absorption in imperfect financial markets' settings (Cohen and Klepper 1996). Note, however, that we explicitly control for access to finance in our regressions. Second, the economies of scale from which large firms benefit allow them to operate with a more efficient division of labor, resulting in better conditions for mechanization and technological upgrade.

Managerial education is strongly positively associated with ISO certification and Internet use. While ISO-certified firms are more likely to be run by older managers, Internet use is more frequent in firms run by younger managers. Firms with a higher share of professionals (engineers, accountants, scientists) are significantly more likely to be ISO certified or to use the Internet. Firms operating with a higher intensity of R&D are also significantly more likely to be ISO certified and to use the Internet. The effects are particularly large for Internet use. The coefficients in column (3) of Table D.5 for the BEEPS 2002 sample imply that (1) an increase in the firm's share of professionals by one standard deviation (22.1 percent) is associated with a 8.3 percent increase in the likelihood of Internet use, whereas (2) an increase in the firm's R&D intensity by one standard deviation (5.6 percent) is associated with a 2.4 percent increase in the likelihood of Internet use.

Our findings provide evidence of skill-biased technological change by showing the importance of complementary investments in skills and R&D for technology absorption. Our estimates suggest that both managerial ability, as well as workforce qualifications, are important determinants of ISO certification and Internet use by firms in ECA. One can claim that professionals are more important than managers in identifying firms'

Table D.6. Determinants of Technology Adoption across Country Groups

| | EU-8 Countries | | | | CIS Countries | | | |
|---|---|---|---|---|---|---|---|---|
| | Dependent Variable Is | | | | Dependent Variable Is | | | |
| | ISO Certification Dummy | | Web Use Dummy | | ISO-Certification Dummy | | Web-Use Dummy | |
| | BEEPS 2002 Sample (1) | BEEPS 2005 Sample (2) | BEEPS 2002 Sample (3) | BEEPS 2005 Sample (4) | BEEPS 2002 Sample (5) | BEEPS 2005 Sample (6) | BEEPS 2002 Sample (7) | BEEPS 2005 Sample (8) |
| **Input markets** | | | | | | | | |
| 50–249 workers | 0.124 [0.030]*** | 0.11 [0.023]*** | 0.1 [0.018]*** | 0.087 [0.012]*** | 0.051 [0.021]** | 0.042 [0.014]*** | 0.208 [0.034]*** | 0.228 [0.024]*** |
| More than 250 workers | 0.23 [0.042]*** | 0.228 [0.043]*** | 0.151 [0.016]*** | 0.081 [0.011]*** | 0.048 [0.026]* | 0.082 [0.024]*** | 0.333 [0.039]*** | 0.336 [0.028]*** |
| Manager with college education dummy | 0.108 [0.033]*** | | 0.234 [0.045]*** | | −0.024 [0.034] | | 0.459 [0.059]*** | |
| Manager age | 0.034 [0.018]* | | 0.148 [0.023]*** | | 0.017 [0.017] | | 0.206 [0.029]*** | |
| Share of professionals | 0.003 [0.001]*** | 0.091 [0.027]*** | −0.001 [0.001] | 0.136 [0.033]*** | 0 [0.001] | 0.049 [0.020]** | −0.007 [0.001]*** | 0.366 [0.045]*** |
| R&D intensity | 0.148 [0.140] | 0.648 [0.366]* | 0.258 [0.192] | 3.813 [1.775]** | 0.103 [0.093] | 0.502 [0.203]** | 0.751 [0.244]*** | 1.372 [0.839] |
| Access to finance dummy | 0.057 [0.016]*** | 0.049 [0.013]*** | 0.098 [0.018]*** | 0.055 [0.011]*** | 0.039 [0.015]*** | 0.019 [0.010]* | 0.078 [0.026]*** | 0.177 [0.019]*** |
| Infrastructure index | −0.027 [0.038] | −0.038 [0.015]** | −0.066 [0.044] | 0.032 [0.013]** | −0.015 [0.007]** | −0.004 [0.006] | 0.112 [0.014]*** | 0.115 [0.012]*** |
| **Market incentives** | | | | | | | | |
| Recently privatized firm dummy | 0.046 [0.037] | −0.061 [0.019]*** | −0.035 [0.047] | −0.004 [0.036] | −0.015 [0.023] | −0.014 [0.018] | 0.033 [0.045] | −0.09 [0.045]** |
| Private firm from origin dummy | 0.008 [0.026] | −0.124 [0.038]*** | 0.003 [0.035] | −0.029 [0.021] | 0.016 [0.020] | −0.005 [0.018] | 0.198 [0.037]*** | 0.037 [0.041] |

106

| | (1) | (2) | (3) | (4) | (5) | (6) | (7) | (8) |
|---|---|---|---|---|---|---|---|---|
| Ownership share of largest shareholder | 0.032 [0.026] | -0.069 [0.022]*** | -0.075 [0.031]** | -0.103 [0.021]*** | -0.043 [0.024]* | -0.011 [0.017] | -0.076 [0.046]* | -0.115 [0.035]*** |
| Dummy for market share less than 5% | -0.082 [0.018]*** | | -0.094 [0.019]*** | | -0.03 [0.016]* | | -0.096 [0.030]*** | |
| Price-cost Margin (%) | | -0.075 [0.055] | | -0.057 [0.034]* | | 0.046 [0.036] | | 0.19 [0.074]** |
| Dummy for pressure to innovate from competitors being important | -0.017 [0.024] | 0.064 [0.014]*** | 0.002 [0.024] | 0.022 [0.016] | 0.007 [0.015] | -0.013 [0.011] | 0.008 [0.027] | 0.036 [0.022] |
| Dummy for pressure to innovate from consumers being important | -0.004 [0.022] | -0.002 [0.021] | 0.029 [0.024] | 0.012 [0.017] | 0.023 [0.014] | 0.021 [0.010]** | 0.028 [0.027] | 0.022 [0.021] |
| Governance index | 0.004 [0.009] | -0.007 [0.007] | 0.007 [0.010] | 0.028 [0.006]*** | -0.005 [0.004] | -0.011 [0.003]*** | -0.041 [0.008]*** | -0.052 [0.007]*** |
| **Access to international knowledge** | | | | | | | | |
| 100% foreign ownership dummy | 0.047 [0.035] | 0.048 [0.036] | 0.083 [0.025]*** | 0.068 [0.014]*** | 0.141 [0.049]*** | 0.055 [0.029]* | 0.447 [0.048]*** | 0.231 [0.043]*** |
| JV dummy (Foreign Ownership < 100%) | 0.02 [0.031] | 0.053 [0.034] | 0.086 [0.028]*** | 0.067 [0.014]*** | 0.057 [0.025]** | 0.027 [0.020] | 0.257 [0.039]*** | 0.132 [0.040]*** |
| Export share | 0.056 [0.033]* | 0.073 [0.029]** | 0.181 [0.055]*** | 0.232 [0.054]*** | 0.067 [0.031]** | 0.054 [0.023]** | 0.314 [0.075]*** | 0.479 [0.074]*** |
| Observations | 1702 | 2490 | 1721 | 2465 | 2391 | 3428 | 2404 | 3441 |

*Notes*: Marginal effects at mean values from probit with random effects regressions shown. Standard errors in parenthesis. ***, **, and * indicates statistical significance at 1%, 5%, and 10% confidence levels, respectively. The regressions include also sectoral dummies, year dummies, and GDP per capita. Higher values of the infrastructure (governance) index indicate better infrastructure (governance).

## Table D.7. Determinants of ISO Certification and Web-Use—Panel Regressions

| | Dependent Variable Is | |
|---|---|---|
| | ISO Certification Dummy | Web Use Dummy |
| | BEEPS Panel Sample | |
| | (1) | (2) |
| **Input markets** | | |
| 50–249 workers | 0.502 [0.114]*** | 0.584 [0.152]*** |
| More than 250 workers | 0.824 [0.147]*** | 1.565 [0.243]*** |
| Share of professionals | 0.809 [0.192]*** | 0.794 [0.240]*** |
| R&D intensity | 1.695 [1.043] | 4.93 [1.335]*** |
| Access to finance dummy | 0.263 [0.089]*** | 0.608 [0.107]*** |
| Infrastructure index | −0.086 [0.040]** | 0.179 [0.046]*** |
| **Market incentives** | | |
| Recently privatized firm dummy | −0.247 [0.176] | −0.246 [0.214] |
| Private firm from origin dummy | −0.053 [0.156] | 0.207 [0.188] |
| Ownership share of largest shareholder | −0.079 [0.156] | −0.735 [0.185]*** |
| Price-cost margin (%) | 0.136 [0.296] | 0.541 [0.346] |
| Dummy for pressure to innovate from competitors being important | 0.109 [0.109] | 0.085 [0.115] |
| Dummy for pressure to innovate from consumers being important | 0.16 [0.100] | 0.178 [0.109] |
| Governance index | −0.052 [0.036] | 0.035 [0.040] |
| **Access to international knowledge** | | |
| 100% foreign ownership dummy | 0.261 [0.164] | 1.239 [0.279]*** |
| JV dummy (foreign ownership < 100%) | 0.532 [0.155]*** | 0.64 [0.223]*** |
| Export share | 0.685 [0.185]*** | 1.482 [0.304]*** |
| Observations | 2310 | 2699 |

*Notes:* Marginal effects at mean values from probit with random effects regressions shown. Standard errors in parenthesis. ***, **, and * indicates statistical significance at 1%, 5%, and 10% confidence levels, respectively. The regressions include also sectoral dummies, year dummies, and GDP per capita. Higher values of the infrastructure (governance) index indicate better infrastructure (governance).

technological problems and solutions. Our findings are also consistent with the absorptive capacity arguments of Cohen and Levinthal (1989): firms invest in R&D not only to pursue innovation directly, but also to "develop and maintain their broader capabilities to assimilate and exploit externally available information." R&D activities are likely to have spillover effects on managerial activities, as firms learn more about their technological bottlenecks and possible solutions through R&D activities, and managers become better informed through their interactions with researchers and engineers. In sum, firms with higher shares of professionals and a higher R&D intensity have a better capacity to process information (to identify problems and solutions). Chapter 3 also finds R&D intensity is systematically associated with technology absorption, yet it makes the important distinction, first made in Cohen and Levinthal (1989), between R&D activities aimed at breakthrough patentable innovations and low-level R&D activities focused on adopting and adapting technology to local conditions. For the vast majority of firms in ECA that do not employ cutting-edge technology, it is the latter type of R&D activities that need to be promoted in order for these firms to catch up to the technological frontier. These activities range from in-house technology adoption and adaptation by a firm's engineers, to the contracting of technological services to local research institutes or individual scientists.

Access to finance is positively and strongly associated with ISO certification and Internet use. Focusing on the results for the BEEPS 2005, a firm with access to a bank loan is 4.2 percent more likely to be ISO certified and 15 percent more likely to use the Internet. Note that access to finance is crucial for firms to acquire new technologies, but may also be important for firms to make the complementary investments needed to better absorb and efficiently use the new technologies. Our findings provide evidence of an important micro channel through which finance may affect growth in ECA—by increasing the absorption of advanced technology by firms.

Better infrastructure (indicated by a higher value of the index) is significantly positively associated with Web use. Since our measure of infrastructure captures the quality of the communications network in the firm's location, this finding is expected, as that quality is fundamental for the Internet technology to be efficiently used by the firm. In contrast, better infrastructure is negatively associated with ISO certification, significantly so in the BEEPS 2002 sample. This result is at odds with the findings for countries such as Brazil in the 2005 Investment Climate Assessment for Brazil. One possible rationale for this counterintuitive sign is that the impact of infrastructure on ISO certification operates through other variables also included in our regressions (for example, foreign ownership or GDP) and as a result, the coefficient on the infrastructure index only partially captures its impact on ISO certification. Another possible rationale is that our infrastructure index does not capture the quality of the road infrastructure. Accounting for the costs of remoteness and the risk of losses in transit would be important to determine the likelihood of ISO certification by a firm. Nevertheless, we should note that our infrastructure index exhibits sensible values, substantially higher for EU-8 and South-eastern European countries, which are characterized by better infrastructure, than for CIS countries or Albania, which are characterized by worse infrastructure.

Our findings on input markets are not driven by any specific subgroup of countries within the ECA region. In fact, we show in Table D.6, the results from estimating our regressions across three country groups: EU-8 countries, CIS countries and Southeastern

Europe and Turkey. The importance of firm size, human capital, R&D intensity, and access to finance for ISO certification and Internet use are verified across all groups of countries. We also find a positive effect of better infrastructure on Internet use and a negative—often weak—effect of better infrastructure on ISO certification across all groups of countries.

## Market Incentives

Next, we discuss the findings on the importance of market incentives. Privatized firms are not more likely—and in the BEEPS 2005 sample are actually significantly less likely—to be ISO certified or to use the Web. Firms that have always been privately owned are not more likely to be ISO certified, but are more likely to use the Internet, and significantly so in the BEEPS 2002 sample. Privatized firms do not seem, therefore, to be the vehicle for productivity gains in ECA countries—which contradicts, to some extent, the recent evidence on productivity improvements due to privatization provided by Brown, Earle, and Telegdy (2007). However, we should note that in unreported regressions that include only the privatization dummy, along with sectoral dummies and GDP per capita, we find that privatized firms were significantly more likely to be ISO certified (in 2002 and 2005), and to use the Internet (in 2005). The fact that these effects disappear or are reversed in the regressions with all factors included, suggests that the advantage of privatized firms in terms of ISO certification or Web use is due to their better input markets, better market incentives, and better access to knowledge.

In unreported regressions where we add an interaction between the privatized dummy and our proxy for concentrated control of ownership shares (discussed below) we find that the coefficient on the interaction term is generally positive, and it is significant in the case of Web use in 2005. This result suggests that privatized firms generate better results (in terms of their access to channels for technology absorption) only when the control issue is solved (when there is a clear private owner with a profit incentive and the power to make changes). However, in Table D.5, the proxy for the concentrated control of ownership shares is significantly negatively associated with ISO certification and Web use. This result is counterintuitive, since firms with better corporate control—for which the concentration of ownership is our proxy—are expected to have better incentives to improve their technology because the profit motive (maximization of the value of the firm) is present, in contrast to the firms with more diffuse ownership.

Firms with smaller market shares in 2002 were significantly less likely to be ISO certified or to use the Internet. Firms charging lower price-cost margins in 2005 were also less likely to be ISO certified or to use the Web, though the effects are weak. Moreover, we find that strong pressure from consumers to innovate increases the likelihood of a firm to be ISO certified or to use the Web. However, the pressure to innovate from competitors has generally no effect on technology absorption, with the exception of a significant positive effect on Internet use in 2005.

In unreported regressions, we estimate the effects on ISO certification and Web use of different measures of market competition, following Carlin, Schaffer, and Seabright (2004). Using a measure of the elasticity of demand faced by the firm from the BEEPS questionnaire, we find that firms facing a more elastic demand (if the firm increased its price, its customers would switch to buying the product from competitors) are less likely to be

ISO certified, or to use the Web. Using a measure based on the total number of competitors a firm faces for its main product in the domestic market, we find no difference in ISO certification and Web use for firms facing no competitors, versus those facing one to three competitors, or more than four competitors in the domestic market.[61] Taken together, our findings suggest that in ECA countries, market concentration is more conducive to ISO certification and Internet use than is competition-which is consistent with the argument in Schumpeter (1942) that competition can be detrimental to innovation, as it reduces the rents that successful innovators can appropriate. As Carlin, Schaffer, and Seabright (2004) argue, firms in ECA face resource constraints that make rents important in financing technology absorption. However, demand pressures generate the need for ECA firms to upgrade their technology.

Better governance (indicated by a higher value of the index) is negatively and significantly associated with ISO certification and Web use. While this result may appear counterintuitive, we attempt to rationalize it as follows. First, this result may reflect reverse causality. Firms that are ISO certified or use the Internet perform better (and are likely to be more profitable), and thus may be more visible to government officials and become more subject to inspections, visits, and bribe extraction, for which proxies enter the calculation of the governance index. Second, the governance result is consistent with the results for competition. In locations where governance is weaker because of poor property rights enforcement and a high potential for expropriation and corruption, the firms that are more able to extract rents are those that have more capabilities to adopt ISO certification, or to use the Internet. While the results on most proxies for market incentives are similar across the three country groups considered in Table D.6, the results for governance differ across country groups. For firms in EU-8 countries better governance is associated with more frequent ISO certification and Web use. This finding is also true to some extent in EU new or candidate countries and the Balkans when it comes to Web use. In EU-8 countries, and to a lesser degree in EU new or candidate countries and the Balkans, better governance provides the right market incentives for ISO certification and Web use. That is not the case in CIS countries, which are driving the negative relationship between governance and ISO certification and Internet use.

### Access to International Knowledge

Finally, we discuss the findings on the importance of access to international knowledge. Relative to domestic firms, firms that are fully foreign owned or joint ventures are more likely to be ISO certified, and especially more likely to use the Internet. These findings are expected, since these firms are embodied in international networks requiring the frequent use of communications technology, and they compete in global markets that require the use of state-of-the-art technological know-how through internationally recognized technical standards. Exporters are also significantly more likely to be ISO certified or to use the Internet. Again, this finding is expected as exporters learn about new technologies through

---

61. The categories of no competitors, one to three competitors, and more than four competitors were those used in the survey to asked the question on competition.

their interaction with more knowledgeable foreign buyers in external markets. While the strong positive association between the participation in export markets and technology adoption is found across the three groups of countries in Table D.6, the effects of foreign ownership are strong only in CIS countries and in EU-accession countries and the Balkans. In EU-8 countries, the advantage of foreign-owned firms relative to domestic firms in technology absorption-related activities is less marked.

The results discussed so far suggest important associations between input markets, market incentives, and access to knowledge on the one hand, and ISO certification or Web use on the other. However, as noted earlier, since the results are based on cross-sections of firms in ECA countries in 2002 and in 2005, they do not allow us to interpret the estimated effects as causal. For example, firms with better managers may be more likely to become ISO certified, but also to hire a larger share of professionals and to participate in export markets. Thus, the positive coefficients on the share of professionals and the export share could simply reflect omitted managerial characteristics.

In order to address the problem that our findings in Table D.5 and Table D.6 could be partly driven by unobservable firm characteristics, we use panel data estimation techniques—probit with random effects—for the BEEPS panel. In this case, we estimate the determinants of ISO certification and Internet use, controlling for unobserved firm invariant characteristics. The results from the panel regressions are shown in Table D.7. The estimated effects of input markets on ISO certification and Internet use are quite similar to those estimated for the 2002 and 2005 cross-sections. Specifically, firms that increase their share of professionals and gain access finance are significantly more likely to be ISO certified, or use the Web. In locations exhibiting better infrastructure, firms are significantly more likely to use the Web, but less likely to be ISO certified. While the effects of R&D intensity on ISO certification are weaker in the panel regressions, they are still very strong for Internet use.

The estimated effects of market incentives on ISO certification and Internet use are somewhat weaker in the panel regressions than in the cross-sectional regressions. We note that firms facing increased pressure from consumers to innovate are significantly more likely to be ISO certified and to use the Web, suggesting demand pull reasons for the firms' technological update.

The estimated effects of access to international knowledge are strong in the panel regressions, and are qualitatively similar to those in the cross-sectional regressions. Foreign ownership and the participation in export markets significantly increase the likelihood of firms being ISO certified and using the Web.

## Summary and Policy Conclusions

This Appendix analyzes the factors that determine ISO certification and Internet use by firms in ECA countries by using the 2002 and 2005 World Bank BEEPS survey dataset. We outline a framework that highlights three types of factors—input markets, market incentives, and access to international knowledge—that may be associated with ISO certification and Internet use. Our findings can guide policy reforms aimed at enabling faster and greater absorption of knowledge and advanced technologies by the private sector. Specifically, we find that firms that have the appropriate *complementary inputs,* namely managerial capacity, higher shares of skilled labor, access to finance (and to a lesser extent access to good infrastructure), and have *access to international knowledge,* either from foreign investors or by exporting, *are more likely to be ISO certified and to use the Web*. Our results suggest that the relationship between market incentives and Internet use or ISO certification is more nuanced. While pressures from consumers generate demand for ECA firms to upgrade their technology, pressures from competitors do not, which may seem counterintuitive in a developed economy setting, but is consistent with previous literature that argues that most firms in ECA face very substantial resource constraints (particularly financial resources), and thus only those with rents are able to finance activities related to technology absorption. Accordingly, we find that large and medium-sized firms are significantly more likely to be ISO certified, and particularly, to use the Web. Our findings also suggest that privatized firms exhibit better ISO certification and Web use outcomes only when there is a clear private owner with a profit incentive and with the power to make changes. As with competition, the negative association we find between governance and ISO certification and Internet use across ECA is counterintuitive; but this effect is very small and driven by the CIS countries. In contrast, in the EU-8 countries, this relationship is reversed, so that better governance is associated with a more business-friendly policy and regulatory regime, and does in fact provide market incentives for ISO certification and for Web use. This indicates that reducing opportunity for rent seeking would raise productive investments in technology absorption. Together with the above-mentioned finding, in environments with severe credit constraints, the firms with substantial rents are more likely to absorb knowledge, and this underscores the importance of improving market incentives *in conjunction with efforts to* improve access to knowledge and critical complementary inputs.

The broad policy implication of these results is that, in order to increase technology absorption by ECA firms, there is a need for *complementary* reforms of the investment climate. Firms should perceive investment in knowledge as the best alternative to increase their value in the long term, and as in any other investment decision, the investment in knowledge should present positive expected net returns. This will require improving firms' access to international knowledge, and increasing the availability of inputs complementary to knowledge within a wider framework of improving market incentives. More specifically, in terms of complementary inputs, our analysis suggests that policy actions in ECA should focus primarily on improving access to credit, in addition to reforming educational systems and supporting lifelong training schemes, so as to improve labor skills and managerial quality. In terms of enhancing access to international knowledge, policymakers should create a favorable environment for local firms to export, and encourage joint ventures between domestic firms and more sophisticated foreign companies. Furthermore, for several years OECD governments have adopted specific programs aimed at addressing the information gaps, transaction costs, and coordination failures in the market for knowledge that are particularly prevalent for SMEs. Such programs have placed strong emphasis on improving firms' capacities to identify technology gaps, and then find and absorb the available technological solution. Other "softer" support measures aimed at diffusing information technology include helping to build private sector-led partnerships or networks between firms, universities, research institutes, and technology transfer centers. As always, the important caveat in any policy intervention is to keep in mind that the final balance between market failures (to be corrected) and government failures (created by the attempt to improve market outcomes), should be such that welfare improvement is expected after the intervention.

# Bibliography

Acemoglu, D., P. Antras, and E. Helpman. 2007. "Contracts and Technology Adoption." *American Economic Review* 97(3):916–43.

Acemoglu, D., S. Johnson, and J. Robinson. 2004. "Institutions as the Fundamental Cause of Long-term Growth." Working Paper No. 10481, National Bureau of Economic Research, Cambridge, Massachusetts.

Aghion P., N. Bloom, R. Blundell, R. Griffith, and P. Howitt. 2005. "Competition and Innovation: An Inverted U Relationship." *Quarterly Journal of Economics* 120(2): 701–28.

Aghion, Philippe, Wendy Carlin, and Mark Schaffer. 2002. "Competition, Innovation and Growth in Transition: Exploring the Interactions between Policies." William Davidson Institute Working Papers Series 501. William Davidson Institute at the University of Michigan Stephen M. Ross Business School, Ann Arbor, Michigan.

Aghion, P., and P. Howitt. 1995. "Technical Progress in the Theory of Economic Growth." In Jean-Paul Fitoussi, ed., *Economics in a Changing World: Proceedings of the 10th World Congress of the International Economic Association*, Volume 5. London: Macmillan.

Aghion, P., and M. Schankerman. 1999. "Competition, Entry and the Social Returns to Infrastructure in Transition Economies." *Economics of Transition* 7(1):79–101.

———. 2004. "On the Welfare Effects and Political Economy of Competition-Enhancing Policies." *Economic Journal* 114(498):800–24.

Aghion, Philippe, Mathias Dewatripont, and Patrick Rey. 1999. "Competition, Financial Discipline and Growth." *Review of Economic Studies* 66(4):825–52.

Aiten, Brian, and Ann E. Harrison. 1999. "Do Domestic Firms Benefit from Direct Foreign Investment: Evidence form Venezuela." *American Economic Review* 89(3):605–18.

Alesina, Alberto, Silvia Ardagna, Giuseppe Nicoletti, and Fabio Schiantarelli. 2005. "Regulation and Investment." *Journal of the European Economic Association* 3(4):791–825.

Alesina, A., and J. Zeira. 2006. "Technology and Labor Regulations." Working Paper 12581, National Bureau of Economic Research, Cambridge, Massachusetts.

Almeida, R., and A. Fernandes. Forthcoming. "Openness and Technological Innovations in Developing Countries: Evidence from Firm-Level Surveys." *Journal of Development Studies*.

Arnold, Jens, Beata S. Javorcik, and Aaditya Mattoo. 2007. "Does Services Liberalization Benefit Manufacturing Firms? Evidence from the Czech Republic." Policy Research Working Paper Series 4109, The World Bank, Washington, D.C.

Arora A., and J. Asundi. 1999. "Quality Certification and the Economics of Contract Software Development: A Study of the Indian Software Service Companies." Working Paper No. 7260, National Bureau of Economic Research, Cambridge, Massachusetts.

Arora, A., and A. Gambardella. 2006. *From Underdogs to Tigers: The Rise and Growth of the Software Industry in Brazil, China, India, Ireland, and Israel*. Oxford University Press.

Athukorola, Prema-Chandra. 2006. "Trade Policy Reform and Structure of Protection in Vietnam." *World Economy* 29(2):161–87.

Baumol, W.J. 1990. "Entrepreneurship: Productive, Unproductive, and Destructive." *Journal of Political Economy* 98(5):893–921.

Beck, T., A. Demirgüç-Kunt, and V. Maksimovic. 2004. "Financial and Legal Constraints to Firm Growth: Does Firm Size Matter?" *Journal of Finance* 60:137–77.

Berdugo, Byniamin, Jacques Sadik, and Nathan Sussman. 2003. "Delays in Technology Adoption, Appropriate Human Capital, Natural Resources and Growth." Available at SSRN: http://ssrn.com/abstract=428142.

Bernard, A.B., and B.J. Jensen. 1995. "Exporters, Jobs, and Wages in U.S. Manufacturing: 1976–1987." *Brookings Papers on Economic Activity* 9(1):5–31.

Bernard, Andrew, Stephen J. Redding, and Peter Schott. 2006. "Multi-Product Firms and Product Switching." Working Papers 12293, National Bureau of Economic Research, Cambridge, Massachusetts.

Blalock, Garrick. 2002. "Technology Adoption from Foreign Direct Investment and Exporting: Evidence from Indonesian Manufacturing." PhD Thesis, University of California, Berkeley.

Blalock, G., and P.J. Gertler. 2004. "Learning from Exporting Revisited in a Less Developed Setting." *Journal of Development Economics* 75(2):397–416.

Blind, K., P. Temple, P. Swann, and G. Williams. 2005. "The Empirical Economics of Standards." DTI Economics Paper 12, UK Department of Trade and Industry, London.

Brahmbhatt, Milan, and Albert Hu. 2007. "Ideas and Innovation in East Asia." Policy Research Paper, WPS4403, The World Bank, Washington, D.C.

Branstetter, Lee. 2006. "Is Foreign Direct Investment a Channel of Knowledge Spillovers: Evidence from Japan's FDI in the United States." *Journal of International Economics* 68:325–44.

Branstetter, Lee, and Yoshiaki Ogura. 2006. Is Academic Science Driving a Surge in Industrial Innovation? Evidence from Patent Citations. Working Paper No. W11561, National Bureau of Economic Research, Cambridge, Massachusetts.

Bridgman, Benjamin R., Igor D. Livshits, and James C. MacGee. 2007. "Vested Interests and Technology Adoption." *Journal of Monetary Economics* 54(3):649–66.

Brown, J., J. Earle, and A. Telegdy. 2006. "The Productivity Effects of Privatization: Longitudinal Estimates from Hungary, Romania, Russia, and Ukraine." *Journal of Political Economy* 114(1):61–99.

Bresnahan, Timothy, and Manuel Trajtenberg. 1995. "General Purpose Technologies: Engines of Growth." Working Paper No. W4148, National Bureau of Economic Research, Cambridge, Massachusetts.

Calderon, Cesar, Norman Loayza, and Luis Serven. 2004. "Greenfield Foreign Direct Investment and Mergers and Acquisitons—Feedback and Macroeconomic Effects." Policy Research Working Paper Series 3192, The World Bank, Washington, D.C.

Carlin Wendy, Mark Schaffer, and Paul Seabright. 2004. "A Minimum of Rivalry: Evidence from Transition Economies on the Importance of Competition for Innovation and Growth," William Davidson Institute Working Papers Series 2004-670, William

Davidson Institute at the University of Michigan Stephen M. Ross Business School, Ann Arbor, Michigan.

Chandra, Vandana. 2006. *Technology, Adaptation, and Exports: How Some Developing Countries Got It Right*. Washington, D.C.: The World Bank.

Chen, Maggie Xiaoyang, Tsunehiro Otsuki, and John S. Wilson. 2006. "Do Standards Matter for Export Success." Policy and Research Working Paper 3809, The World Bank, Washington, D.C.

Clerides, S, S. Lach, and J. Tybout. 1998. "Is Learning-by-Exporting Important? Micro-Dynamic Evidence from Columbia, Mexico and Morroco." *Quarterly Journal of Economics* 113(3):903–47.

Coe, David T., and Elhanan Helpman. 1995. "International R&D Spillovers." *European Economic Review* 39:859–87.

Cohen, D., P. Garibaldi, and S. Scarpetta. 2004. *The ICT Revolution: Productivity Differences and the Digital Divide*. Oxford: Oxford University Press.

Cohen, W., and S. Klepper. 1996. "Firm Size and the Nature of Innovation Within Industries: The Case of Process and Product R&D." *Review of Economics and Statistics* 78(2):232–43.

Cohen, W., and R. Levin. 1989. "Empirical Studies of Innovation and Market Structure." In R. Schmalensee and R. Willig, eds., *Handbook of Industrial Organization*, Vol. II. Elsevier Science Publishers.

Cohen, W., and D. Levinthal. 1989. "Innovation and Learning: The Two Faces of R&D." *Economic Journal* 99(397):569–96.

———. 1990. "A New Perspective on Learning and Innovation." In special issue on "Technology, Organizations, and Innovations," *Administrative Science Quarterly* 35(1):128–52.

Commander, Simon, and Janos Kollo. 2004. "The Changing Demand for Skills: Evidence from the Transition." IZA DP No. 1073. Institute for the Study of Labor (IZA), Bonn.

Commander, Simon, and Jan Svejnar. 2007. "Do Institutions, Ownership, Exporting and Competition Explain Firm Performance? Evidence from 26 Transition Countries." IZA DP No. 2637. Institute for the Study of Labor (IZA), Bonn.

Conway, Paul, Donato De Rosa, Giuseppe Nicoletti, and Faye Steiner. 2006. "Regulation, Competition and Productivity Convergence." Working Paper No. 509, OECD Economics Department, Paris.

Corbett, C., M. Montes-Sancho, and D. Kirsch. 2005. "The Financial Impact of ISO 9000 Certification in the United States: An Empirical Analysis." *Management Science* 51(7):1046–59.

Correa, P, A.M. Fernandes, and C. Uregian. 2008. "Technology Adoption and the Investment Climate: Firm-Level Evidence for Eastern Europe and Central Asia."

Dodok, R., D. Glisic, D. Jovanovic, M. Popovic, and E. Romhanji. 2006. "Effect of Annealing Temperature on the Formability of Al–Mg4.5–Cu0.5 Alloy Sheets." Proceedings of the 11th International Conference on Metal Forming. *Journal of Materials Processing Technology* 177(1-3):386–89.

Department of Trade and Industry. 2005. "The Empirical Economics of Standards." DTI Economics Paper 12, UK Department of Trade and Industry, London.

De Loecker, Jan. 2007. "Do Exports Generate Higher Produtivity? Evidence from Slovenia." *Journal of International Economics* 73(1):69–98.

Demirbag, M., R. Huggins, and V. Ratcheva. 2006. "Foreign Direct Investment Flows: Patterns Across the Globe." Academy of International Business Conference, UK Chapter, Manchester, UK.

Demirbag, M., R, Huggins, and V. Ratcheva. 2007. "Global Knowledge and R&D Foreign Direct Investment Flows: Recent Patterns in Asia Pacific, Europe and North America." *International Review of Applied Economics* 21(3):437–51.

Desai, Raj M., and Itzhak Goldberg. 2001. "The Politics of Russian Enterprise Reform: Insiders, Local Governments, and the Obstacles to Restructuring." *World Bank Research Observer* 16:219–40.

———. 2008 (forthcoming). "*Can Russia Compete: Enhancing Productivity and Innovation in a Globalizing World.*" Washington, D.C.: Brookings Press.

Djankov, Simeon, and Bernard M. Hoekman. 2000. "Foreign Investment and Productivity Growth in Czech Enterprises." *World Bank Economic Review* 14(1):49–64.

Djankov, Simeon, and Peter Murrell. 2002. "Enterprise Restructuring: A Quantitative Survey." *Journal of Economic Literature* 40(3):739–92.

Earle, John, and Saul Estrin. 2003. "Privatization, Competition, and Budget Constraints: Disciplining Enterprises in Russia." *Economics of Planning* 36(1):1–22.

Eaton, J., and S. Kortum. 2001. "Trade in Capital Goods." *European Economic Review* 45(7):1195–1235.

———. 2002. "Technology, Geography, and Trade." *Econometrica* 70(5):1741–79.

Eisenhardt, Kathleen M. 1989. "Building Theories From Case Study Research." *The Academy of Management Review* 14(4):532.

Engman, Michael. 2005. "The Economic Impact of Trade Facilitation." Working Paper No. 21. Available at: www.sourceoecd.org/10.1787/861403066656, Organization for Economic Co-operation and Development, Paris.

Ernst, Dieter. 2004. "Pathways to Innovation in the Global Network Economy: Asian Upgrading Strategies in the Electronics Industry." Working Papers 58, East-West Center, Economics Study Area. Hawaii.

Ernst, Dieter, and Linsu Kim, 2002. "Global Production Networks, Knowledge Diffusion, and Local Capability Formation." *Research Policy* 31(8-9):1417–29.

Eschenbach, Felix, and Bernard Hoekman. 2005. "Services Policy Reform and Economic Growth in Transition Economies, 1990–2004." Policy Research Working Paper 3663, The World Bank, Washington, D.C.

Esfahani, H.S., and M.T. Ramirez. 2003. "Institutions, Infrastructure, and Economic Growth." *Journal of Development Economics* 70(2):443–77.

Estrin, S., X. Richet, and J. Brada, eds. 2000. *Foreign Direct Investment in Central Eastern Europe: Case Studies of Firms in Transition.* M.E. Sharpe, Inc.

European Commission. 2007. "Key Figures 2007: Towards a European Research Area—Science, Technology, and Innovation." Brussels.

Feenstra, R., and H. Kee. 2004. "On the Measurement of Product Variety in Trade." *American Economic Review* 94:145–49.

Feenstra, R., D. Madani, T.H. Yang, and C.Y. Liang. 1999., Testing Endogenous Growth in South Korea and Taiwan. *Journal of Development Economics* 60:317–41.

Fernandes, A.M. 2007. "Structure and Performance of the Services Sector in Transition Economies." Policy Research Working Paper 4357, The World Bank, Washington, D.C.

Girma, S., and Y. Gong. Forthcoming "FDI, Linkages and the Efficiency of State-owned Enterprises in China." *Journal of Development Studies.*

Goldberg, Itzhak. 2004. *Poland and the Knowledge Economy—Enhancing Poland's Competitiveness in the European Union.* Washington, DC: The World Bank.

Goldberg, Itzhak, and John Nellis. Forthcoming. "Methods and Institutions—How Do They Matter: Lessons From Privatization and Restructuring In The Post-Socialist Transition." In Ira Lieberman and Daniel Kopf, eds., *Privatization in Transition Economies: The Ongoing Story.* New York: JAI Press, Elsevier Science, Inc.

Goldberg, Itzhak, and Branko Radulovic. 2005. "Hard Budget Constraints, Restructuring and Privatization in Serbia: A Strategy for Growth of the Enterprise Sector." Private Sector Note, The World Bank.

Goldberg, Itzhak, Branko Radulovic, and Mark Schaffer. 2005. "Productivity, Ownership and the Investment Climate: International Lessons for Priorities in Serbia." Policy Research Working Paper 3681, The World Bank, Washington, D.C.

Goldberg, Itzhak, Manuel Trajtenberg, Adam Jaffe, Thomas Muller, Julie Sunderland, and Enrique Blanco Armas. 2006. "Public Financial Support for Commercial Innovation: Europe and Central Asia Knowledge Economy Study Part I." Chief Economist's Regional Working Paper Series 1 (1), The World Bank, Washington, D.C.

Griliches, Z., Hybrid Corn, and the Economics of Innovation. 1957. "An Exploration in the Economics of Technological Change." *Econometrica* 25:501–22.

Grossman, Gene M., and Elhanan Helpman. 1991. "Trade, Knowledge Spillovers, and Growth." Working Paper No. W3485. National Bureau of Economic Research, Cambridge, Massachusetts.

Griffith. R., S. Redding, and J. Van Reenen. 2004. "Mapping the Two Faces of R&D: Productivity Growth in A Panel of OECD Industries." *Review of Economics and Statistics* 86(4):883–95.

Grossman, G., and E. Helpman. 1991. *Innovation and Growth in the Global Economy.* Cambridge, Mass.: MIT Press.

Haddad, Mona, Jaime de Melo, and Brendan Horton. 1996. "Morocco, 1984–89: Trade Liberalization, Exports and Industrial Performance." In Mark J. Roberts and James R. Tybout, eds., *Industrial Evolution in Developing Countries: Micro Patters of Turnover, Productivity, and Market Structure.* Oxford: Oxford University Press.

Hall, B.H, A.B. Jaffe, and M. Trajtenberg. 2000. "Patent Citations and Market Value: A First Look." Working Paper No. 7741, National Bureau of Economic Research, Cambridge, Mass.

Hall, Bronwyn H., Adam B. Jaffe, and Manuel Trajtenberg. 2001. "The NBER Patent Citations Data File: Lessons, Insights, and Methodological Tools." Discussion Paper 3094, Centre for Economic Policy Research (C.E.P.R), London.

Hausmann, Ricardo, and Dani Rodrik. 2003. "Economic Development As Self-Discovery." *Journal of Development Economics* 72(2):603–33.

Harrison, Ann. 1996. "Determinants and Effects of Direct Foreign Investment in Côte d'Ivoire, Morocco, and Venezuela." In Mark J. Roberts and James R. Tybout, eds., *Industrial Evolution in Developing Countries: Micro Patters of Turnover, Productivity, and Market Structure.* Oxford: Oxford University Press.

Helpman, Elhanan, and Manuel Trajtenberg. 1996. "Diffusion of General Purpose Technologies." Working Paper No W5773, National Bureau of Economic Research. Cambridge, Mass.

Henderson, R., A.B. Jaffe, and M. Tratenberg. 1999. "Universities as a Source of Commerical Technology: A Detailed Analysis of University Patenting 1965–1988." Working Papers 5068, National Bureau of Economic Research, Cambridge, Mass.

Hoekman, Bernard and Javorcik, Beata. 2006. *Global Integration and Technology Transfer*. Washington, D.C.: The World Bank.

Humphrey, J., and H. Schmitz. 2000. "Governance and Upgrading: Linking Industrial Cluster and Global Value Chain Research." Working Paper 120, Institute of Development Studies, University of Sussex, Brighton, UK.

Javorcik, Beata. 2004. "Does Foreign Direct Investment Increase the Productivity of Domestic Firms? In Search of Spillovers through Backward Linkages." *American Economic Review* 94(3):605–27.

Jensen, Jesper, Thomas Rutherford, and David Tarr. 2007. "The Impact of Liberalizing Barriers to Foreign Direct Investment in Services: The Case of Russian Accession to the World Trade Organization." *Review of Development Economics* 11(3):482–506.

Jensen, Jesper, and David Tarr. Forthcoming. "Impact of Local Content Restrictions and Barriers against Foreign Direct Investment in Services: The Case of Kazakhstan Accession to the WTO." *Eastern European Economics*.

Jones, Ronald, Henryk Kierzkowski, and Chen Lurong. 2004. "What Does the Evidence Tell us about Fragmentation and Outsourcing." Working Paper No: 09/2004, The Graduate Institute of International Studies (HEI), Geneva, Switzerland.

Jorgenson, D. 2001. "Information Technology and the U.S. Economy." *American Economic Review* 91(1):1–32.

Javorcik, Beata, Wolfgang Keller, and James Tybout. 2006. "Openness and Industrial Response in a Wal-Mart World: A Case Study of Mexican Soaps, Detergents and Surfactant Producers." Policy Research Paper 3999. The World Bank, Washington, D.C.

Kee, Hiau Looi, Alessandro Nicita, and Marcelo Olarreaga. 2006. "Estimating Trade Restrictiveness Indices." Policy Research Working Paper 3840, The World Bank, Washington, D.C.

Keller, W. 1996. "Absorptive Capacity: On the Creation and Acquisition of Technology in Development." *Journal of Development Economics* 49:199–227.

Keller, W., and S.R. Yeaple. 2003. "Multinational Enterprises, International Trade and Productivity Growth: A Firm Level Evidence from the United States." Working Paper No. 9504, National Bureau of Economic Research, Cambridge, Mass.

Keller, Wolfgang .1998. "Are International R&D Spillovers Trade-Related? Analyzing Spillovers among Randomly Matched Trade Partners." *European Economic Review* 42(8):1469–81.

———. 2000. "Do Trade Patterns and Technology Flows Affect Productivity Growth?" *The World Bank Economic Review* 14(1):17–47.

———. 2002a. "Geographic Localization of International Technology Diffusion." *American Economic Review* 92(1):120–42.

———. 2002b. "Trade and the Transmission of Technology." *Journal of Economic Growth* 7(1):5–24.

King, R., and R. Levine. 1993. "Finance and Growth: Schumpeter Might be Right." *Quarterly Journal of Economics* 108(3):717–37.

Konings, Jozef. 2000. "The Effects of Foreign Direct Investment on Domestic Firms: Evidence from Firm Level Panel Data in Emerging Economies." Working Papers Series 344, William Davidson Institute at the University of Michigan Stephen M. Ross Business School, Ann Arbor, Michigan.

Kraay, Aart. 1999. "Exportations et Performances Economiques: Etude d'un Panel d'Enterprises Chinoises." *Revue d'Economie Du Developpement* 1-2:183–207.

Kraay, Aart. Isidro Soloaga, and James Tybout. 2002. "Product Quality, Productive Efficiency, and International Technology Diffusion: Evidence from Plant-level Panel Data." Policy Research Working Paper Series 2759, The World Bank, Washington, D.C.

Krug, Jeffrey A., and W. Harvey Hegarty. "Predicting Who Stays and Leaves after an Acquisition: A Study of Top Managers in Multinational Firms." *Strategic Management Journal* 22:185–96.

Krugman, Paul. 1995. "Technology, Trade, and Factor Prices." Working Paper 5355, National Bureau of Economic Research, Cambridge, Mass.

Lall, Sanjaya. 1984. "India's Technological Capacity: Effects of Trade, Industrial and Science and Technology Policies." In M. Fransman and K. King, eds., *Technological Capability in the Third World*. London: Macmillan.

———. 1999. "Multinationals and Technology Development in Host LDCs." In J. Cantwell, ed., *Foreign Direct Investment and Technological Change, Vol. 2*. Cheltenham, UK: Edward Elgar Publishing.

Lieberman, Ira, and Daniel Kopf, eds. 2007. *"Privatization in Transition Economies: The Ongoing Story."* New York: JAI Press, Elsevier Science Inc.

Love, I., and R. Gatti. 2006. "Does Access to Credit Improve Productivity? Evidence from Bulgarian Firms." Policy Research Working Paper 3921, The World Bank, Washington, D.C.

Lucas, R. 1988. "On the Mechanics of Economic Development" *Journal of Monetary Economics* 22(1):3–42.

Maskus, Keith E., Tsunehiro Otsuki, and John S. Wilson. 2005. "The Cost of Compliance with Product Standards for Firms in Developing Countries: An Econometric Study." Policy and Research Working Paper 3560, The World Bank, Washington, D.C.

Mattoo, Aaditya. 2005. "Economics and Law of Trade in Services." Policy and Research Working Report, The World Bank, Washington, D.C.

Mattoo, Aaditya, Randeep Rathindran, and Arvind Subramanian. 2001. "Measuring Services Trade Liberalization and its Impact on Economic Growth: An Illustration." Policy Research Working Paper No. 2655, The World Bank, Washington, D.C.

Moran T.H., E.D. Graham, and M. Blomström, eds. 2005. *Does Foreign Direct Investment Promote Development?* Washington, D.C.: Institute for International Economics.

Navaretti, G.B., I. Soloaga, and W. Takacs. 1998. "When Vintage Technology Makes Sense: Matching Imports to Skills." Policy Research Working Paper Series 1923, The World Bank, Washington, D.C.

Nelson, R., and E. Phelps. 1966. "Investment in Humans, Technological Diffusion, and Economic Growth." *American Economic Review* 56(1/2):65–75.

Nicoletti, Giuseppe, and Stefano Scarpetta. 2003. "Regulation, Productivity and Growth: OECD Evidence." *Economic Policy* 18(36):9–72.

Nicoletti, Giuseppe, and Stefano Scarpetta. 2005. "Regulation and Economic Performance: Product Market Reforms and Productivity in the OECD." Economics Department Working Papers, No.460, OECD, Paris.

North, Douglass C. 1991. "Institutions." *Journal of Economic Perspectives* 5(1):97–112.

Pack, Howard. 1993. "Technology Gaps between Industrial and Developing Countries: Are There Dividends for Late-comers?" Proceedings of the World Bank Annual Conference on Development Economics, The World Bank, Washington, D.C.

Pack, Howard, and Kamal Saggi. 2001. "Vertical Technology Transfer via International Outsourcing." *Journal of Development Economics* 65(2):289–415.

Pack, Howard, and Larry E. Westphal. 1986. "Industrial Strategy and Technological Change." *Journal of Development Economics* 22:87–128.

Parente, Stephen L., and Edward C. Prescott. 1994. "Barriers to Technology Adoption and Development." *Journal of Political Economy* 102(2):298–321.

———. 1999. "Monopoly Rights: A Barrier to Riches." *American Economic Review* 89(5):1216–33.

Pavlova, A. 2001. "Adjustment Costs, Learning-by-doing, and Technology Adoption under Uncertainty." Working Paper 4369-01, MIT Sloan School of Management, Boston.

Pavcnik, Nina. 2002. "Trade Liberalization, Exit and Productivity Improvements: Evidence from Chilean Plants." *Review of Economic Studies* 69:245–76.

Persson, Maria. 2007. "Trade Facilitation and the EU-ACP Economic Partnership Agreements: Who Has the Most to Gain?" Working Paper Series No. 2007:8, Lund University, Department of Economics, Sweden.

Roberts, M., and J. Tybout. 1997. "An Empirical Model of Sunk Costs and the Decision to Export." *American Economic Review* 87(4):515–61.

Rodrik, Dani. 2004. "Industrial Policy For The Twenty-First Century." Paper No. 4767, Centre for Economic Policy Research (C.E.P.R), London.

Roller, L., and L. Waverman. 2001. "Telecommunications Infrastructure and Economic Development: A Simultaneous Approach." *American Economic Review* 91(4):909-23.

Romer, Paul. 1986. "Increasing Returns and Long-run Growth." *Journal of Political Economy* 94(5):1002–37.

Romhanji, E., M. Popovic, D. Glisic, R. Dodok, and D. Jovanovic. 2006. "Effect of Annealing Temperature on the Formability of Al-Mg4.5-Cu0.5 Alloy Sheets." Proceedings of the 11th International Conference on Metal Forming. *Journal of Materials Processing Technology* 177(1-3):386–89.

Rutherford, Thomas, David Tarr, and Oleksandr Shepotylo. Forthcoming. "Poverty Effects of Russia's WTO Accession: Modeling 'Real Households' with Endogenous Productivity Effects." *Journal of International Economics.*

Shepherd, Ben, and John S. Wilson. 2006. "Road Infrastructure in Europe and Central Asia: Does Network Quality Affect Trade?" Policy ResearchWorking Paper No. 4104, The World Bank, Washington, D.C.

Shepotylo, Oleksandr, and David Tarr. Forthcoming. "Specific Tariffs, Tariff Simplification, and the Structure of Import Tariffs in Russia: 2001–2005." *Eastern European Economics.*

Schiff, Maurice, Yanling Wang, and Marcelo Olarreaga. 2002. "Trade-related Technology Diffusion and the Dynamics of North-South and South-South Integration." Policy and Research Working Paper 2861, The World Bank, Washington, D.C.

Schumpeter, J. 1942. *Capitalism, Socialism, and Democracy*. New York: Harper.

Solow, Robert. 1956. "A Contribution to the Theory of Economic Growth." *Quarterly Journal of Economics* 70(1):65–94.

Stiroh, K. 2002. "Information Technology and the U.S. Productivity Revival: What do the Industry Data Say?" *American Economic Review* 92(5):1559–76.

Sutton, Robert I., and Anita L. Callahan. 1987. "The Stigma of Bankruptcy: Spoiled Organizational Image and Its Management." *The Academy of Management Review* 30(3):405–36.

Tan, Savchenko, Gimpelson, Kapelyushnikov, Lukyanova, Kalashnikov. 2007. "Skills Shortages and Training in Russian Enterprises." The World Bank, Washington, D.C.

Van Biesebroeck, Johannes. 2005. "Exporting Raises Productivity in Sub-Saharan African Manufacturing Firms." *Journal of International Economics* 67(2):373–91.

Van Biesebroeck, Johannes. 2006. "The Sensitivity of Productivity Estimates: Revisiting Three Important Productivity Debates." *Journal of Business and Economic Statistics*.

Walsh, James P., and John W. Elwood. 1991. "Mergers, Acquisitions, and the Pruning of Managerial Deadwood." *Strategic Management Journal* 12:201–17.

Watkins, Alfred, Isak Frouman, and Irina Dezhina. 2006. *From Knowledge to Wealth: Integrating Science and Higher Education for the Development of Russia*. Washington, D.C.: The World Bank.

Wilson, John S., Catherine L. Mann, and Tsunehiro Otsuki. 2003. "Trade Facilitation and Economic Development: A New Approach to Quantifying the Impact." *World Bank Economic Review* 17(3):367–89.

Wilson, John, and Tsunehiro Otsuki. 2001. "Global Trade and Food Safety—Winners and Losers in a Fragmented System." Policy and Research Working Paper 2689, The World Bank, Washington, D.C.

World Bank. 2004. "Creating a Better Investment Climate for Everyone." *World Development Report*. Washington, D.C.

———. 2005a. *From Disintegration to Reintegration: Eastern Europe and Former Soviet Union in International Trade*. Washington, D.C.

———. 2005b. "Hard Budget Constraints, Restructuring and Privatization in Serbia: A Strategy for Growth of the Enterprise Sector Private Sector Note." Private Sector Note, The World Bank, Washington, D.C.

———. 2005c. *Investment Climate Assessment Brazil*. Washington, D.C.

———. 2007a. *The Path to Prosperity: Productivity Growth in Eastern Europe and the Former Soviet Union*. Washington, DC: World Bank.

———. 2007b. *Productivity Growth, Job Creation, and Demographic Change in Eastern Europe and the Former Soviet Union*. Washington, D.C.

———. 2008. *Global Economic Prospects 2008: Trends in Technology Diffusion in Developing Countries*. Washington, D.C.

Zeira, J. 1998. "Workers, Machines, and Economic Growth." *The Quarterly Journal of Economics* 113(4):1091–1117.

# Eco-Audit

## Environmental Benefits Statement

The World Bank is committed to preserving Endangered Forests and natural resources. We print World Bank Working Papers and Country Studies on 100 percent postconsumer recycled paper, processed chlorine free. The World Bank has formally agreed to follow the recommended standards for paper usage set by Green Press Initiative—a nonprofit program supporting publishers in using fiber that is not sourced from Endangered Forests. For more information, visit www.greenpressinitiative.org.

In 2007, the printing of these books on recycled paper saved the following:

| Trees* | Solid Waste | Water | Net Greenhouse Gases | Total Energy |
|---|---|---|---|---|
| 264 | 12,419 | 96,126 | 23,289 | 184 mil. |
| '40' in height and 6–8" in diameter | Pounds | Gallons | Pounds $CO_2$ Equivalent | BTUs |

green press INITIATIVE